NOW YOU KNOW!

BY JOSHUA FREE

COMMUNICATION, CONTROL & BETA-COMMAND PROFESSIONAL COURSE
A GRADE-IV MARDUKITE SYSTEMOLOGY PUBLICATION

NOW YOU KNOW!

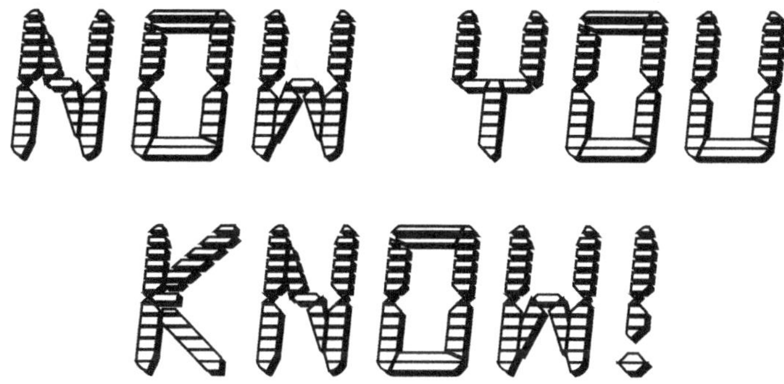

THE SECRET OF UNIVERSES & HOW YOU GOT STUCK IN ONE

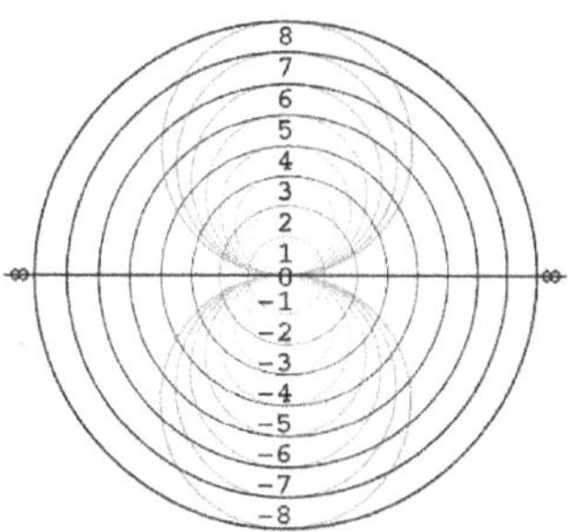

WRITTEN BY JOSHUA FREE
with contributions and
FOREWORD BY DAVID ZIBERT

FIRST EDITION—JULY 2020

© 2020, JOSHUA FREE

AN OFFICIAL MARDUKITE SYSTEMOLOGY PUBLICATION

Based on the training materials developed by Joshua Free
for publication in the Research and Discovery Series of
the *Mardukite Research Organization, Mardukite Zuism*
and *NexGen Systemology Society* (and its affiliates).

Original Underground Title: "Help One Another" (June 2020)
Mardukite Research Library Catalogue No. "Liber-3C"

International School of Systemology
Systemology Grade-IV : Metahuman Systemology

Systemology logos and cover graphic assistance by Kyra Kaos
Interior images excerpted from lecture-charts by Joshua Free

mardukite.com

COMMUNICATION, CONTROL AND COMMAND OF BETA-EXISTENCE
GRADE-IV MARDUKITE SYSTEMOLOGY PROFESSIONAL COURSE

We all "know" we are "eternal spirits" entrapped into having these "human experiences"—so, *what are you going to do about it?*

Many esoteric models of universal cosmology and spiritual ascension appear on the timeline of recorded Human history over the past 6,000 years, and the most ancient of these records—the *Arcane Tablets*—reveal the most simple and complete account of Cosmic History, that which later inspired an entire planet of cultural mythologies, traditions and religious interpretations. Yet none of these further fragments and facets of the original *Crystal* ever brought a clearer experience or more perfect understanding than what had come before. As a result, the truth inherent in the simplicity once shared became forgotten and lost to a sea of "symbols" and "representations" that reflected the poorest shadows of a former age—and the *Ancient Mystery School* was born.

What has now developed underground over the course of a decade for Mardukite Zuism and Systemology far surpasses any attempts made by former paradigms—each one reaching further and further away from its original source, never returning the Mind-System and its perfect control by the Spirit, back to the original state that it once maintained before its fragmentation into the various human systems.

We know that other universes exist and that we have occupied higher existences, but now we have forgotten the way out. And perhaps for the first time, in a long time, there is a recognizable way out; solutions revealed to us now in the 21st century A.D. of what has only been touched upon in obscurity since the original cuneiform tablet renderings of the 21st century B.C.

The most effective corrective measure we have found is systematically running all of this programming and its circuits *backwards*, and are directing Seekers "back the way they came" and not imposing some new artificial course of action. The decay or pattern-cycle of time is merely relative to the consideration of events registered as having taken place along a spiritual timeline.

Back behind the considerations assumed and energetically imprinted on a personal track, the true Self—the Alpha Spirit—is still there, bright, beautiful and powerful; the "I" that *is* the *Awareness* of the individual... if you can but remember what you chose to forget...

We are ready to face these challenges now... *Are you?*

Before starting this present course you may wish to complete
the first part of this Grade-IV material previously released as
"Communication and Control of Energy & Power" (Vol.1)
and also "Command of the Mind–Body Connection" (Vol.2)

COMMUNICATION, CONTROL & BETA-COMMAND—TABLET OF CONTENTS
"NOW YOU KNOW!"—THE TRUTH ABOUT UNIVERSES & HOW YOU GOT STUCK IN ONE

:: Foreword ::
THOUGHT PROCESSES, THOUGHTFORMS
& GATES OF MENTAL FRAGMENTATION
[by David Zibert]

Although not exactly encouraged in our society—or formally trained in any way—I would like to tell you a bit about why it is not only interesting, but rather important, for you to inquire about the Mind; its various fragmentation and thought processes as a system. What I'm referring to here as "The Mind," is the human faculty of "reason," or else "Awareness"—which, in part, differentiates humanity—and the Human Condition—from all other life forms.

Surely enough, all life has its own capacity for reasoning—within the limitation of its own species or genetic vehicle[1]—its own "Mind-System" you could say; be it animal, insects, plants, even minerals are alive in their own ways.[‡] But there is definitely something that sets the Human Condition[2] apart from all of these other forms of life—and that's what I am referring to here as the "Mind-System." It is our grand peculiarity as a species...

That being said: we sure share a lot—have a lot—in common with other life forms; and there is no particular problem with that—we are "All-as-One" here—but if this "Mind" we are talking about is what sets us apart: *what* is unique to *our* current experience of life on Earth? Maybe that's worth exploring a bit; worth inquiry to know about. Maybe it is worth spending some time contemplating and understanding—as human beings—what we have in common with, for example, dogs. Not to be completely overlooked or dismissed, we might appreciate dogs more when living life as—or extending our point-of-view as—a dog, don't you think?

> The bottom line is: command of the "Mind-System" is the key to a superior experience of life and reality.

When you're "thinking up" something, functions of the "Mind-System" appear to be quite straightforward. You just "do it"—and either ideas and/or memories seem to simply pop up in your head; often in words for more abstract concepts, but also perceived as images, sounds, even smells or tastes. This is especially true when it comes to memory recall; since

---

1 **genetic-vehicle** : a physical *Life*-form; the physical (*beta*) body that is animated/controlled by the (*Alpha*) *Spirit* using a continuous *Lifeline* (ZU); a physical (*beta*) organic receptacle and catalyst for the (*Alpha*) *Self* to operate "causes" and experience "effects" within the *Physical Universe*

‡ See also Joshua Free's edition of "*Pantheisticon: The Wizard Handbook of John Toland*"—also reprinted in the Grade-I "Route of Druidism & Dragon Legacy" Master Edition anthology titled: "*Merlyn's Complete Book of Druidism: A Master Course in Druidry for Modern Druids*" by Joshua Free.

2 **Human Condition** : a standard default state of Human experience that is generally accepted to be the extent of its potential identity (*beingness*)—currently treated as *Homo Sapiens Sapiens,* but which is scheduled for replacement by *Homo Novus.*

all information from the senses can be associated[3] and imprinted in the ideas and mental images too.

What you might be unaware of, is that there is a systematic process going on in within and as your "Mind"—and, of course, biochemically taking place throughout your brain as you "think up" stuff. This is all part of what I refer to as the "Thought Process." And certainly let us not confuse ourselves here—the "Mind" is not the "brain." In this instance, the brain is a biological machine used to process activities of the "Mind-System" for a genetic vehicle in this Physical Universe. The brain is meat; the mind is energy.

Δ Δ Δ Δ Δ Δ Δ

THE THOUGHT PROCESS... When you "think up" something—and I say "something," as "thoughts are things" in their own right. When you think up something, where does it come from? Basically, from nowhere—or what can be described as the "unmanifest reality," or else "infinite potentiality."[4] In effect, *thinking* is simply a means to communicate with reality— solidifying it. Just thinking a thought instantly manifests it within the reality of your mental universe, engaging the "Mind-System," which is always comparing and evaluating[5] thoughts and data with previous experience[6] and former thoughts in order to have it "make sense."

Your Mind-System is always "on" and always computing right, correctly within itself; essentially incapable of a wrong computation. All of the errors come from erroneous[7] data supplied and collected through experiences; the analytical functions of the Mind as an operating system is not what is at fault. Even when you do understand that your calculations are wrong concerning some subject or effort, the same processes and faculties[8] apply, as the mind corrects itself to be right in the face of more "accurate" experience and data.

Every thought you have gets stored in some folder within the Mind-System, creating your chain of memory, associated together to create artificial structures and patterns—and each

---

3   **associative knowledge** : significance or meaning of a facet or aspect assigned to (or considered to have) a direct relationship with another facet; to connect or relate ideas or facets of existence with one another; a reactive-response image, emotion or conception that is suggested by (or directly accompanies) something other than itself; in traditional systems logic, an equivalency of significance or meaning between facets or sets that are grouped together, such as in *(a + b) + c = a + (b + c)*; in NexGen Systemology, erroneous associative knowledge is assignment of the same value to all facets or parts considered as related (even when they are not actually so), such as in *a = a, b = a, c = a* and so forth without distinction.

4   **potentiality** : the total "sum" (collective amount) of "latent" (dormant—present but not apparent) capable or possible realizations; used to describe a state or condition of what has not yet manifested, but which can be influenced and predicted based on observed patterns and, if referring to beta-existence, Cosmic Law.

5   **evaluate** : to determine, assign or fix a set value, amount or meaning.

6   **experiential data** : accumulated reference points we store as memory concerning our "experience" with Reality.

7   **erroneous** : inaccurate; incorrect; containing error.

8   **faculties** : abilities of the mind (individual) inherent or developed.

subsequent thought is compared to—and "processed" through—each file in each associated folder, which further validates the classification and evaluation of these patterns as memory.

The Thought Process operates, as I previously mentioned, like a communication[9] system with reality—your own personal reality universe and that of others; and the same thoughts and memories might be filed differently in different individuals[10] and given a different consideration. That's why, for example, different individuals might recall different facets[11] from the same event in a unique way. In our Physical Universe, we tend to think of communication largely based on "words" or "semantics"[12]—each composing our complete self-confirming paradigm[13] for understanding[14] reality and communications of the same, such as is seen as much in the academic community as in the esoteric[15] occult underground.

Paradigms and semantic systems are all approximations of reality. Each contributes to the creation of biased "thought-forms"[16] ultimately leading the individual to a biased experience of reality—with the same erroneous data stored for all future evaluations of reality perception. These fixed paradigms—operating in exclusion to Self-Honesty[17]—are artificial,

---

9 **communication** : successful transmission of information, data, energy (&tc.) along a message line, with a reception of feedback; an energetic flow of intention to cause an effect (or duplication) at a distance; the personal energy moved or acted upon by will or else 'selective directed attention'; the 'messenger action' used to transmit and receive energy across a medium.

10 **individual** : a person, lifeform or human entity; a *Seeker* or potential *Seeker* is often referred to as an "individual" within Mardukite Zuism and Systemology materials.

11 **facets** : an aspect, an apparent phase; one of many faces of something; a cut surface on a gem or crystal; in *NexGen Systemology*—a single perception or aspect of a memory or "*Imprint*"; any one of many ways in which a memory is recorded; perceptions associated with a painful emotional (sensation) experience and "*imprinted*" onto a metaphoric lens through which to view future similar experiences; other secondary terminals that are associated with a particular terminal, painful event or experience of loss, and which may exhibit the same *encoded* significance as the *activating event*.

12 **semantics** : the *meaning* carried in *language* as the *truth* of a "thing" represented, *A-for-A*; the *effect* of language on *thought* activity in the Mind and physical behavior; language as *symbols* used to represent a concept, "thing" or "solid."

13 **paradigm** : an all-encompassing *standard* by which to view the world and *communicate* Reality; a standard model of reality-systems used by the Mind to filter, organize and interpret experience of Reality.

14 **understanding** : a clear 'A-for-A' duplication of a communication as 'knowledge', which may be comprehended and retained with its significance assigned in relation to other 'knowledge' treated as a 'significant understanding'; the "grade" or "level" that a knowledge base is collected and the manner in which the data is organized and evaluated.

15 **esoteric** : hidden; secret; knowledge understood by a select few.

16 **thought-form** : apparent *manifestation* or existential *realization* of *Thought-waves* as "solids" even when only apparent in Reality-agreements of the Observer; the treatment of *Thought-waves* as permanent *imprints* obscuring *Self-Honest Clarity* of *Awareness* when reinforced by emotional experience as actualized "thought-formed solids" ("*beliefs*") in the Mind.

17 **Self-honesty** : the *alpha* state of *being* and *knowing*; clear and present total *Awareness* of-and-as *Self*, in its most basic and true proactive expression of itself as *Spirit* or *I-AM*—free of artificial attachments, perceptive filters and other emotionally-reactive or mentally-conditioned programming imposed on the human condition by the systematized physical world.

first consciously created and then maintained as an automatic mechanism that is entirely corruptible as per Natural Law[18] or even the entropy[19] taking place in this Physical Universe.

The true origin of all personal thought is Self—and the energy comes from an Infinite Source and not from lower-level[20] machinery restricted to the material senses. When one is confined to a mental point-of-view within the reactionary level of the physical body, the "brain" may indeed operate as a physical machine that relays orders to other control centers related to the "body" as a genetic vehicle, but these are all reactive stimulus-response mechanisms—and they may lead to false conclusions about reality when allowed to dictate the associative thought in the Mind-System. But, as previously stated, the brain is not the mind.

It might be the case that some stuff in the universe that seems very real and valid to you fits nowhere into the agreed[21] upon paradigm of consensual[22] reality, the mutually held beliefs about the objective "Physical Universe." When thoughts run contrary to these "beliefs," the understanding is deemed non-sensical or insane. Thus, we end up with all of these differing worldviews throughout the timeline[23] of history when humans did not "compute" well with one another, and even in the broadcast of history to the present, when modern humans are out of communication with the past, likely to consider earlier civilizations as idiotic and non-sensical; hence why most of its true secrets remained buried for so long.

One might notice how some individuals have a tendency to invoke accepted authorities[24] in order to validate their own agreed-upon thought-data—such as we see within the realm of accepted academia—the contemporary academia of any age—where promoting original ideas is not only frowned upon, but mostly forbidden, whether by the "laws" of specially funded sciences or by organized religions. Both have their coffers to fill. The irony here being that these alleged "accepted" authorities go through the same thought process and are subject to

---

18 **Cosmic Law** : the "Law" of Nature (or the Physical Universe); the "Law" governing cosmic ordering; often called "Natural Law" in sciences and philosophies that attempt to codify or systematize it.

19 **entropy** : the reduction of organized physical systems back into chaos-continuity when their integrity is measured against space over time.

20 **level** : a physical or conceptual *tier* (or plane) relative to a *scale* above and below it; a significant *gradient* observable as a *foundation* (or surface) built upon and subsequent to other levels of a totality or whole; a *set* of "*parameters*" with respect to other such *sets* along a *continuum*; in *NexGen Systemology*, a *Seeker's* understanding, *Awareness* as *Self* and the formal grades of material/instruction are all treated as "*levels.*"

21 **agreement** : unanimity of opinion; an accepted arrangement; "reality."

22 **consensual** : formed or existing simply by consent—by general or mutual agreement; permitted, approved or agreed upon by majority of opinion; knowingly agreed upon unanimously by all concerned; to be in agreement on the objective universe and/or a course of action therein.

23 **timeline** : plotting out history in a linear (line) model to indicate instances (experiences) or demonstrate changes in state (space) as measured over time; a singular conception of continuation of observed time as marked by event-intervals and changes in energy and matter across space.

24 **authoritarian** : knowledge as truth, boundaries and freedoms dictated to an individual by a perceived, regulated or enforced "authority."

the same fragmentation[25] herein described in Joshua Free's "*Liber-3C*." Consensual reality is indeed both malleable and corruptible—and whomever controls the most agreed upon paradigm controls that reality.

If you are seeking an actual Self-Honest reality experience, the way is to free yourself and be able to think beyond the paradigms and semantic systems you have been formerly programmed with. Some esoteric mystics have called this "Crossing to the Abyss,"[26] or else "Antinomian Thinking."[27] But whatever name it is given, the core of our Systemology toward Self-Honesty is not a *new* discovery; for it has been known in select underground circles for a very long time. It was, for example, blatantly known to the early Christian Gnostics[28]—and is also reflected in even older historical texts such as the "*Chaldean Oracles*" or the Babylonian[29] "*Epic of Creation.*"[*]

Δ Δ Δ Δ Δ Δ Δ

AN EXAMPLE OF ONE SUCH PARADIGM... the QBL or "*Kabbalah*"—as developed through Rabbinical Jewish lore from knowledge first concealed within the "Gate" system of Babylon—is perhaps one of the more widely known mystical paradigms employed in Western[30] esoterica. I will therefore use it here as an introductory example, for those that are familiar with it, to illustrate the semantics of the Mind-System Thought Process. The semantics from

---

25 **fragmentation** : breaking into parts and scattering the pieces; the *fractioning* of wholeness or the *fracture* of a holistic interconnected *alpha* state, favoring observational *Awareness* of perceived connectivity between parts; *discontinuity*; separation of a totality into parts; in *NexGen Systemology*, a person outside a state of *Self-Honesty* is said to be *fragmented*.

26 **Crossing the Abyss** : to enter the spiritual or metaphysical unknown in "Self-annihilation" to purify the Self and "return to the Source."

27 **antinomian** : a term applied to *Gnostics* (popularized by Martin Luther during the Christian reformation) denoting a rejection of formal religious morals and dogma—decreed, written and interpreted by humanity—as a true pathway to Ascension (some elements appear in all forms of religious protest and reformation but as an extreme, would be considered spirto-religious rebellious punkdom by some modern standards, but it should be understood that it does follow a higher ethic, such as Mardukite Utilitarianism.

28 **Gnostics** : a name meaning "having knowledge" in Greek language (see also *gnosis*); an early sect of Judeo-Christian mysticism from the 1st Century AD emphasizing true knowledge by *Self-Honest* experience of metahuman and spiritual states of beingness, emphasizing defragmentation of "illusion" and the overcoming of material "deception"; an esoteric proto-Systemology organization disbanded by the Roman Church as heretical.

29 **Babylonian** : the Mesopotamian civilization that evolved from *Sumer*; the inception of all societal and religious systematization.

* See materials contained within "*The Complete Anunnaki Bible: A Source Book of Esoteric Archaeology*" edited by Joshua Free—also available within the complete Mardukite *Grade-II* Master Edition hardcover anthology titled: "*Necronomicon: The Complete Anunnaki Legacy.*"

30 **Western Civilization** : the modern history, culture, ideals, values and technology, particularly of Europe and North America as distinguished by growing urbanization and industrialization and born from a rebellion to strong religious indoctrination.

the Semitic Kabbalah are not directly a part of Mardukite Zuism[31] & Systemology, but they are a way in which some Seeker's[32] will already be familiar with the concepts explored within and therefore tend to be reviewed in lower Grades.[∞]

When you "think up" something, it comes from AIN, the sea of infinite potentialities, then it manifests[33] in the reality of Kether, the "Crown," and then *swoosh!* The whole "Tree of Life" manifests rather instantly in response, and the thought is computed to fit somewhere on this systematized model—largely based on associative data within a semantic paradigm—to make the content and evaluations of the thought compute with reality, preventing the individual's universe from collapsing on itself. And, of course, several different semantics and interpretations for the same Qabbalistic model exist to even fit *that* knowing into some predisposed category of knowing.

You, therefore, unknowingly—or even knowingly—place each and every thought and memory upon one or another "Sphere of Existence"—the *Sephiroth* in the QBL—and these get fixed and stored in your memory bank. And when thinking or recalling something else, the Mind-System takes into account everything that is already stored in this memory bank as a comparison in order to file the thought or memory once again into a "Sphere of Existence" and fix it upon a timeline, confirming it and having it make sense, thus once more preventing the mind from collapsing.

In this sense, thoughts confirm and feed other thoughts endlessly; keeping you busy in this mind maze, never letting you see the actual thought or memory as it truly is in a defragmented[34] state, outside of a paradigm system or other erroneous associations. This mental fragmentation is what defines how your own personal reality or universe is measured against other realities or universes, cementing to create the agreed upon, or else consensual reality—*this* one you are experiencing right now in *this* Physical Universe—as they manifest in the Sphere of Malkuth, again using the QBL as an example. But we could just as well apply

---

31 **Mardukite Zuism** : a Mesopotamian-themed (Babylonian-oriented) religious philosophy and tradition applying the spiritual technology based on *Arcane Tablets* in combination with "Tech" from *NexGen Systemology*; first developed in the New Age underground by Joshua Free in 2008 and realized publicly in 2009 with the formal establishment of the "*Mardukite Chamberlains.*"

32 **Seeker** : an individual on the *Pathway to Self-Honesty*; a practitioner of *Mardukite Systemology* or *NexGen Systemology Processing* that is working toward *Ascension*.

∞ See *Mardukite Liber-K,* released in the *"Practical Babylonian Magic"* paperback anthology by Joshua Free and the hardcover edition, "*Necronomicon—The Anunnaki Grimoire*" (also contained within the complete *Grade-II* Master Edition hardcover "*Necronomicon: The Complete Anunnaki Legacy.*"

33 **manifestation** : something brought into existence.

34 **defragmentation** : the *reparation* of wholeness; a process of removing "*fragmentation*" in data or knowledge to provide a clear understanding; applying techniques and processes that promote a *holistic* interconnected *alpha* state, favoring observational *Awareness* of continuity in all spiritual and physical systems; in *NexGen Systemology*, a "*Seeker*" achieving an actualized state of basic "*Self-Honest Awareness*" is said to be *defragmented*.

this to any version of the "Tree of Life" or "World Tree" or "Chakra Centers"[35] or "Gate-System." They are all semantic paradigms for relaying the Thought Processes of the Mind-System toward manifestation of an enforced[36] reality, or else the "condensation of universes" as will be explored more specifically in this present volume, "*Liber-3C.*"

Now, in regards to an individual trying to get out of this process: you can, of course, attempt to shift paradigms all you want; and this is pretty much what most of the mystical occult scene—and even the "New Thought" movement—is presently concerned with; with endless presentations of personal "gnosis,"[37] offering only a different paradigm for a different experience of reality semantics. But these, while interesting, have a tendency to only fragment and obscure one's understanding of reality and their personal universe even further. Hence these other paths are really akin to simply getting out of a cage just to enter a more alluring one. Do you see that?

What we have developed over the course of a decade for Mardukite Zuism and Systemology far surpasses the attempts made by these former paradigms—each one reaching further and further away from its original source, never returning the Mind and its control back to the original state that it once maintained before its fragmentation into the various human systems. And perhaps for the first time, in a long time, there is a recognizable way out; solutions revealed to us now in the 21st century A.D. of what has only been touched upon in obscurity since the original cuneiform tablet renderings of the 21st century B.C.

We are ready to face these challenges now... *Are you?*

    —David Zibert
      A.T. Lab Office, Quebec, Canada
      July 1, 2020

---

35 **chakra** : (an archaic term used by ancient wisdom traditions); an etheric wheel-mechanism that processes *ZU* energy at specific frequencies along the *ZU-line*, of which a Human being reportedly has *seven* at various degrees.

36 **enforcement** : the act of compelling or putting (effort) into force; to compel or impose obedience by force; to impress strongly with applications of stress to demand agreement or validation; the lowest-level of direct control by physical effort or threat of punishment; a low-level method of control in the absence of true communication

37 **gnosis** : a *Greek* word meaning knowledge, but specifically "true knowledge"; the highest echelon of "true knowledge" accessible (or attained) only by mystical or spiritual faculties whereby actualized realizations are achieved independent of specialized education.

:: Introduction ::

THIS IS SYSTEMOLOGY—A HISTORY & OVERVIEW
« GUIDE TO THE GRADE-III ESOTERIC LIBRARY »
[Summation Presented by David Zibert[*]]

Since the inception[38] of the NexGen Systemology Society[39] nearly a decade ago, many things have been brewing quietly and unseen in the underground; but, fear not, as slowly but surely, everything will be brought to light...as "New Babylon" *is rising.*

The original literary presentation of NexGen Systemology occurred underground in 2011 and continued through 2013. The essential materials from this period were recently reissued as *"Systemology: The Original Thesis"* by Joshua Free. These materials first began to appear in 2011 with a series of booklets by Joshua Free, which at first glance were actually quite different from anything he had really presented before. The booklets were the first to present "Systemology"—or else, the work of the "Systemological Society" as an offshoot of the Mardukite Research Organization and extension of the Mardukite Chamberlains group that previously participated in the development of the Grade-II Mardukite Core.[∞]

> Of course, "Systemology" stands for "system logics"—or else, "the logics behind the systems," which is also to say, in more esoteric terms, "the magic behind the magic."

In 2011, several booklets were released in the "original thesis" Systemology series; the first titled *"Human, More Than Human: Awakening to the Next Evolution."* It was really a down to earth approach; a really simple user-friendly booklet about how, quite literally, "Humans are more than Human"—that we are more than simply our physical body and that there are really "worlds" out there that most individuals are unaware of in their daily lives. It was really just taking the reader by the hands and saying in a rather basic and gentle way how *we are more than human.*

The original underground release and presentation of Systemology was quite peculiar at the time. Even I wasn't sure what the goal was behind all this. But in the end, it made sense—and there was a brief follow-up published soon after: *"Systemology Defragmentation: Self-Honesty for the Next Evolution."* This title delved into the core of the matter and explained the basic

---

[*]   Based on a transcript of two video presentations delivered to the "Mardukite Chamberlains" and "NexGen Systemology Society" by David Zibert in November 2019.

38 **inception** : the beginning, start, origin or outset.

39 **NexGen Systemology** : a modern tradition of applied religious philosophy and spiritual technology based on *Arcane Tablets* in combination with *"general systemology"* and *"games theory"* developed in the New Age underground by Joshua Free in 2011 as an advanced futurist extension of the *"Mardukite Chamberlains."*

∞   The Grade-II Esoteric Research Library or "Mardukite Core" is collected in its entirety within *"Necronomicon: The Complete Anunnaki Legacy"* (2020 Master Edition Hardcover) by Joshua Free.

"defragmentation" process, which is the same as the ascent[40] up the "Ladder of Light" as we know it from the *Grade-II* presentation and the more commonly known Babylonian paradigm.

Systemology presented the core pathway; the same pathway intended with the previous "Mardukite Core" and our explorations into the Babylonian paradigm proper, but these new booklets presented the main tenets of this core without some of the more esoteric, magical or religious semantic trappings that we commonly find with literal interpretations of the *Arcane Tablets*.

A third booklet comprising the "original thesis" arrived in 2012 as *"Transhuman Generations: The Next Evolution of a Species."* This one relates how worldviews are programmed in the generational cycles that repeat itself over an over. And, of course, when most individuals are unaware of that taking place, such as we see in the world that we live in, history is bound to repeat itself. Although the material is quite basic, it is really important information for Self-Actualization.[41]

Another installment, completing the "original thesis," appeared in 2013, titled *"Systemology For Life: Patterns and Cycles."* This one continued in the spirit of *"Transhuman Generations,"* but this relayed material about personal cycles—and about cycles repeating themselves—yes, through the generations, but more specifically, cycles repeating themselves as we experience them as individuals: how to notice them, and to go beyond them, of course, toward the goal of *Self-Honesty* which is, again, achieved via "defragmentation."[42]

It is these materials—that I've just mentioned—that are gathered together into the small anthology reissued now as *"Systemology: The Original Thesis."*[∞] In addition to these, there were a few other small underground releases that were not as widely circulated. One of these being *"The Games: Portals of Self-Transformation & The Underground Occult Initiation."*[*] This was a very controversial booklet when it was published; it related the adventure of Joshua Free in the West Coast Occult Underground.

---

40 **ascension** : actualized *Awareness* elevated to the point of true "spiritual existence" exterior to *beta existence*. An "Ascended Master" is one who has returned to an incarnation on Earth as an inherently *Enlightened One*, demonstrable in their actions—they have the ability to *Self-direct* the "Spirit" as *Self*, just as we are treating the "Mind" and "Body" at this current grade of instruction; in *Moroii ad Vitam*, a state of Beingness after *First Death*, experienced by an *etheric body*, which is able to maintain consciousness as a personal identity continuum with the same *Self-directed* control and communication of Will-Intention that is exercised, actualized and developed deliberately during one's present incarnation.

41 **actualization** : to make actual; to bring into Reality; to realize fully in *Awareness*.

42 **defragmentation** : the *reparation* of wholeness; a process of removing *"fragmentation"* in data or knowledge to provide a clear understanding; applying techniques and processes that promote a *holistic* interconnected *alpha* state, favoring observational *Awareness* of continuity in all spiritual and physical systems; in *NexGen Systemology*, a "Seeker" achieving an actualized state of basic *"Self-Honest Awareness"* is said to be *defragmented*.

∞ Also available in the complete Grade-III Master Edition hardcover *"Systemology Handbook."*

* Portions of this text are reprinted in the complete Grade-III *"Systemology Handbook."*

It also became apparent in 2013 that many individuals, even those among the Mardukite network still studying the *Grade-II* "Mardukite Core," were not ready for Joshua Free's new "Systemology" developments as a whole. Very few outside the small network of the "NexGen Systemological Society" really took notice of what we working toward from 2011 to 2013—and continued to work even more quietly and unnoticed thereafter. Of course, this was all about to change with a reboot that is presently going on now as we enter the 2020's.

One of the very interesting aspects of "Mardukite Systemology" is experienced by one who has been actually following the work of Joshua Free from the beginning—and seeing how "Systemology" *was* the goal; seeing how it was the *goal* all along since the very beginning and the inception of the Mardukite Ministries and the Mardukite Research Organization in 2008. Anyone who has read the introductory material of the "*Necronomicon: The Anunnaki Bible*" or "*The Complete Anunnaki Bible*" by Joshua Free—even those individuals that have read his *Arcanum* book—will notice that the "logic of systems" and an aim toward the applied spiritual technology of a "Systemology" is what is underlying the journey and is what has been there all along, driving the work forward.

In October 2019, the NexGen Systemology Society experienced a new public debut with the arrival of its first true core textbook, catalogued[43] as *Mardukite Systemology "Liber-One,"* released globally as "*The Tablets of Destiny: Using Ancient Wisdom to Unlock Human Potential*"[‡] from the new Joshua Free Publishing Imprint. It concisely presents the fundamental foundation for "Mardukite Systemology" itself—upon which further Systemology discourses are now built upon.

"*The Tablets of Destiny*" (*Liber-One*) is *not* a rehash of what was done before (with "*Systemology: The Original Thesis*")—it is actually a completely new presentation. It presents the logics behind the systems, of what the "Mardukite Chamberlains" had discovered concerning Babylon. It's like: "Okay, we found *that*. Now, *what* do we *do* with it?—And how does everyone get *benefit* from it?" Now, this is where we are. This is what we do. So, now what is this "*Grade-III Mardukite Systemology*" you may ask?

At the most basic core: it is an applied spiritual technology of the 21st century AD based on the spiritual wisdom from the 21st century BC, which were compiled in their rawest tablet form in the *Grade-II* Mardukite Core—"*The Complete Anunnaki Bible*," "*Sumerian Legacy*," &tc. Launching *Grade-III*, *Liber-One* introduces what we have termed the "Standard Model"[44] (of

---

43 **catalog** : a systematic list of knowledge or record of data.

‡ Also available in the complete Grade-III "*Systemology Handbook.*"

44 **standard model** : a fundamental *structure* or symbolic construct used to evaluate a complete *set* in *continuity* relative to itself and variable to all other *dynamic systems* as graphed or calculated by *logic*; in *NexGen Systemology*—our existential and cosmological cabbalistic model; a "*monistic continuity model*" demonstrating *total system* interconnectivity "above" and "below" observation of any apparent *parameters*; the *ZU-line* represented as a singular vertical (*y*-axis) waveform in space across dimensional levels (universes) without charting any specific movement across a dimensional time-graph *x*-axis.

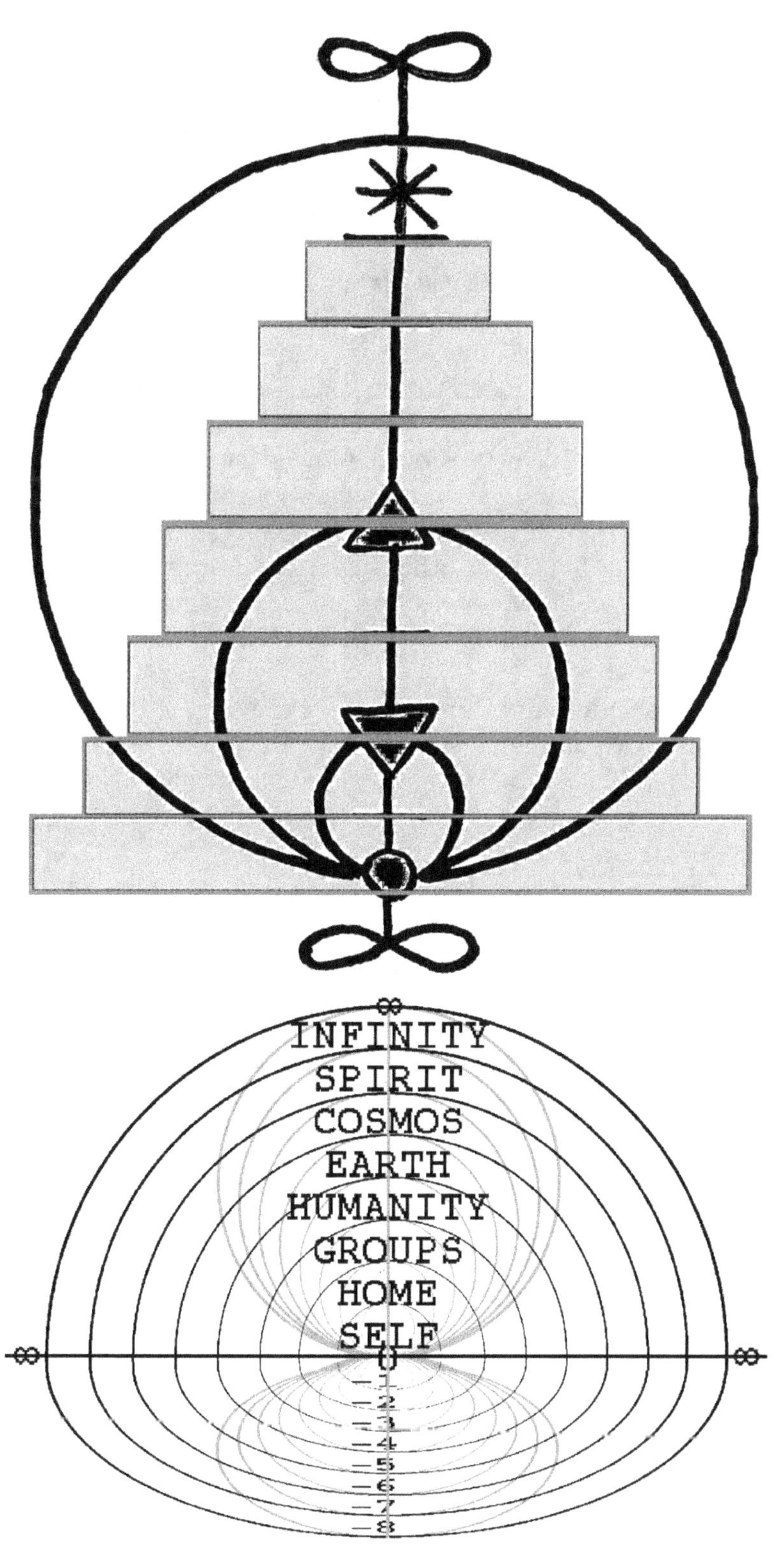

∞
INFINITY
SPIRIT
COSMOS
EARTH
HUMANITY
GROUPS
HOME
SELF
∞
∞
0
1
2
3
4
5
6
7
8

Systemology), otherwise known as the "*ZU-Line*"[45] (in Mardukite Zuism). This, in itself, is a workable non-dogmatic and applied model of the same Babili Ladder of Lights—the StarGates of this Universe—of which you should already be familiar with from *Grade-II* work.

The "Standard Model" or "ZU-Line"—*how does it work?*

First of all, the Standard Model or *ZU-Line* is an abstract construct, graphically defining parameters[46] for a Systemology of the Human Condition. It is divided as *seven*—or *eight*—steps for practical purposes, but theoretically extends to Infinity, above and below its scale; just like the Ladder of Lights paradigm of *Gates*, or any such similar Kabbalistic Model.

"*The Tablets of Destiny*" (*Liber-One*) focuses on the lower levels of the scale—from 0-to-4. This work emphasizes building a strong personal foundation of health and strength before the individual is introduced to more advanced practices—such as those included in its follow up manual, "*Crystal Clear*" (*Liber-2B*) and the other upper-level Grades. But, most importantly, we found out that a sane "Mind-Body Connection" is a prerequisite to experiencing a Self-Honest, clear and unfragmented realization[47] of Self as "I-AM-Alpha-Spirit"[48] in this lifetime.

This new approach is actually quite different from previous attempts in other traditions; even the most pious *Gnostic* paradigms still continue to reject material existence[49]—what we

---

45 **Zu-line** : a spectrum of *Spiritual Life Energy (ZU)* as conceived on the Standard Model of Systemology; an energetic channel of *Identity-continuum* connecting the *Awareness (ZU)* of an *Alpha-Spirit* with "*Infinity*"; a *Life-line* on which *Awareness (ZU)* extends from the direction of the "Spiritual Universe" (AN) as its *alpha state* through an entire possible range of activity in its *beta state*, experienced as a *genetic-entity* occupying the *Physical Universe (KI)*; the Standard Model demonstrates the Zu-line interacting with spheres of existence.

46 **parameters** : a defined range of possible variables within a model, spectrum or continuum; the extent of communicable reach capable within a system or across a distance; the defined or imposed limitations placed on a system or the functions within a system; the extent to which a Life or "thing" can *be*, *do* or *know* along any channel within the confines of a specific system or spectrum of existence.

47 **realization** : the clear perception of an understanding; to make "real" or give "reality" to so as to grant a property of "beingness" or "being as it is"; the state or instance of coming to an *Awareness*; in *NexGen Systemology*, "gnosis" or true knowledge achieved during *systematic processing*; achievement of a new (or "higher") cognition, true knowledge or perception of Self in relation to reality.

48 **alpha-spirit** : a "spiritual" *Life*-form; the "true" *Self* or I-AM; the spiritual (*alpha*) *Self* that is animating the (*beta*) physical body or "*genetic vehicle*" using a continuous *Lifeline* of spiritual ("*ZU*") energy; an individual spiritual (*alpha*) entity possessing no physical mass or measurable waveform (motion) in the Physical Universe as itself, so it animates the (*beta*) physical body or "*genetic vehicle*" as a catalyst to experience *Self*-determined causality in effect within the *Physical Universe*.

49 **existence** : the *state* or fact of *apparent manifestation*; the resulting combination of the Principles of Manifestation: consciousness, motion and substance; continued *survival*; that which independently exists; the '*Prime Directive*' and sole purpose of all manifestation or Reality; the highest common intended motivation driving any "*Thing*" or *Life*.

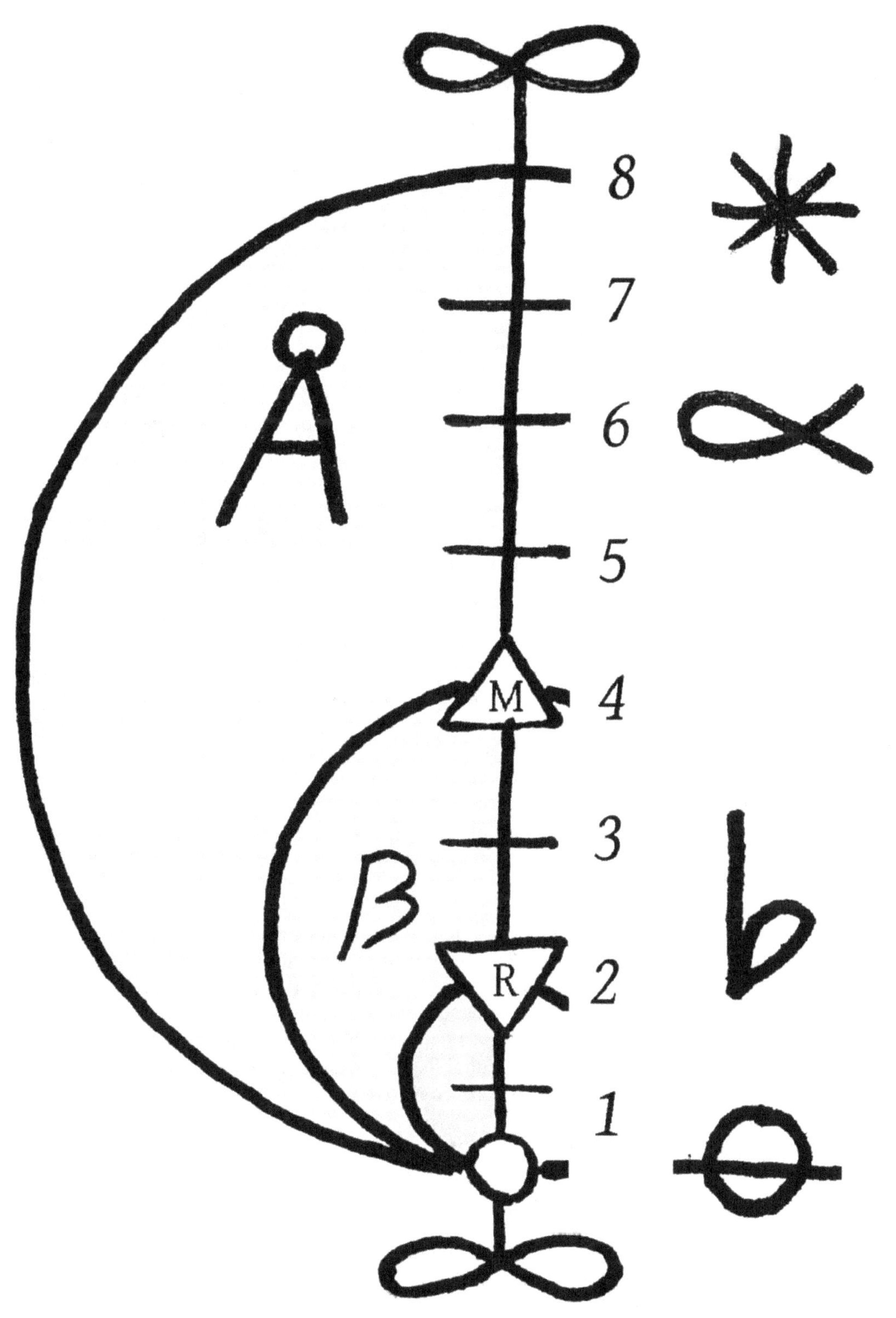

8
7
6
5
4
M
3
B
R
2
1
A

refer to as "beta-existence"[50]—as an "illusion." We are not rejecting the Physical Universe in Mardukite Systemology; no, rather we are agreeing that such is an artificial ordination of an otherwise very real universe in which the Human being, as a "genetic vehicle,"[51] is the tool used to experience of such a reality. Lower gradients[52] of the ZU-line (Standard Model) run as follows—but, be aware that these descriptions are something of an over simplification; there is more to it, though this should suffice for our present review:

0 — Inert Matter (theoretical zero, since everything in existence is basically a motion), or else Body Death (for the "genetic vehicle");

1 — Physical Body (basic physiological[53] functions/cellular "fight/flight") receiving communications from...

2 — Reactive Control Center (or "RCC")[54] which includes survival programming ("reactive response mechanism") inherent to the development and experience of all physical life.

Between "1" and "2" lies the standard emotional range of the Human Condition, which can be, and is, programmed and encoded with *imprints*[55] preventing the Self access to its own experience of higher levels in Self-Honesty.

---

50 **beta (existence)** : all manifestation in the "Physical Universe" (KI); the "Physical" state of existence consisting of vibrations of physical energy and physical matter moving through physical space and experienced as "time"; the conditions of *Awareness* for the *Alpha-spirit* (*Self*) as a physical organic *Lifeform* or "*genetic vehicle*" in which it experiences causality in the *Physical Universe*.

51 **genetic-vehicle** : a physical *Life*-form; the physical (*beta*) body that is animated/controlled by the (*Alpha*) *Spirit* using a continuous *Lifeline* (ZU); a physical (*beta*) organic receptacle and catalyst for the (*Alpha*) *Self* to operate "causes" and experience "effects" within the *Physical Universe*.

52 **gradient** : a degree of partitioned ascent or descent along some scale, elevation or incline; "higher" and "lower" values in relation to one another.

53 **physiology** : a science of motions of living bodies or organisms.

54 **reactive control center (RCC)** : the secondary (reactive) communication system of the "*Mind*"; a relay point of *Awareness* along the Identity's *ZU-line*, which is responsible for engaging basic motors, biochemical processes and any *programmed automated responses* of a living *beta* organism; the reactive Mind-Center of a living organism relaying communications of *Awareness* between causal experience of *Physical Systems* and the "*Master Control Center*"; it presumably stores all emotional encoded imprints as fragmentation of "chakra" frequencies of *ZU* (within the range of the "*psychological/emotive systems*" of a being), which it may *react* to as Reality at any time; in *NexGen Systemology*, this is plotted at (2.0) on the continuity model of the *ZU-line*.

55 **imprint** : to strongly impress, stamp, mark (or outline) onto a softer 'impressible' substance; to mark with pressure onto a surface; in *NexGen Systemology*, the term is used to indicate permanent Reality impressions marked by frequencies, energies or interactions experienced during periods of emotional distress, pain, unconsciousness, loss, enforcement, or something antagonistic to physical (personal) survival, all of which are are stored with other reactive response-mechanisms at lower-levels of *Awareness* as opposed to the active memory database and proactive processing center of the Mind; an experiential "memory-set" that may later resurface—be triggered or stimulated artificially—as Reality, of which similar responses will be engaged automatically.

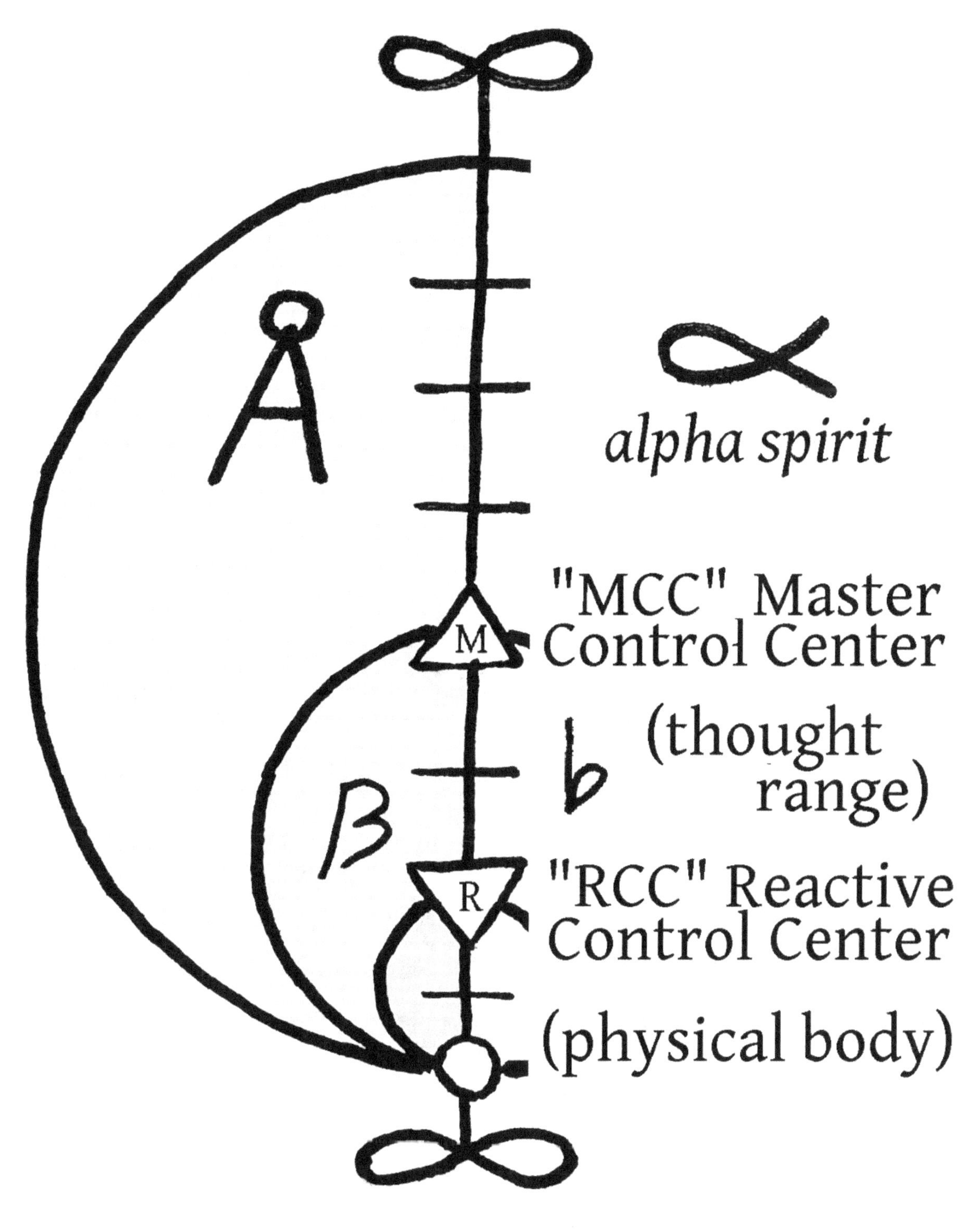

A
alpha spirit
"MCC" Master Control Center
(thought range)
"RCC" Reactive Control Center
(physical body)

3 — Thought (associative knowledge)[56] and activity communicated from...

4 — Master Control Center (or "MCC")[57] which is the point of contact from the True Self, or Higher Self in some paradigms, and is the highest gradient relating to the genetic vehicle or physical body for the Human Condition in the Physical Universe.

Self-Honesty[58] is to be sought at each of these gradients as one moves upward on the "Pathway." This means that, for example, if you are not at a point of Self-Honesty regarding gradient "1" and "2," then you won't be certain to have a Self-Honest command of the thoughts and programming beyond that, which is preventing you from experiencing the knowing and beingness of your Higher or True Self; which we refer to as the state of the "Alpha Spirit," a spiritual being which merely maintains considerations[59] of experiencing a beta-existence.

In effect what is sought on this "Pathway" is a clear communication[60] with the continuity[61] of All Life—and that starts with your own—and when you have a Self-Honest experience of your

---

56 **associative knowledge** : significance or meaning of a facet or aspect assigned to (or considered to have) a direct relationship with another facet; to connect or relate ideas or facets of existence with one another; a reactive-response image, emotion or conception that is suggested by (or directly accompanies) something other than itself; in traditional systems logic, an equivalency of significance or meaning between facets or sets that are grouped together, such as in $(a + b) + c = a + (b + c)$; in NexGen Systemology, erroneous associative knowledge is assignment of the same value to all facets or parts considered as related (even when they are not actually so), such as in $a = a$, $b = a$, $c = a$ and so forth without distinction.

57 **master control center** (MCC) : a perfect computing device to the extent of the information received from "lower levels" of sensory experience/perception; the proactive communication system of the "*Mind*"; a relay point of active *Awareness* along the Identity's *ZU-line*, which is responsible for maintaining basic *Self-Honest Clarity* of *Knowingness* as a *seat of consciousness* between the *Alpha-Spirit* and the secondary "*Reactive Control Center*" of a *Lifeform* in *beta existence*; the Mind-center for an *Alpha-Spirit* to actualize cause in the *beta existence*; the analytical *Self-Determined* Mind-center of an *Alpha-Spirit used* to project *Will* toward the genetic body; the point of contact between *Spiritual Systems* and the *beta existence*; presumably the "*Third Eye*" of a being connected directly to the *I-AM-Self*, which is responsible for *determining* Reality at any time; in *NexGen Systemology*, this is plotted at (4.0) on the continuity model of the *ZU-line*.

58 **Self-honesty** : the *alpha* state of *being* and *knowing*; clear and present total *Awareness* of-and-as *Self*, in its most basic and true proactive expression of itself as *Spirit* or *I-AM*—free of artificial attachments, perceptive filters and other emotionally-reactive or mentally-conditioned programming imposed on the human condition by the systematized physical world.

59 **consideration** : careful analytical reflection of all aspects; deliberation; determining the significance of a "thing" in relation to similarity or dissimilarity to other "things"; evaluation of facts and importance of certain facts; thorough examination of all aspects related to, or important for, making a decision; the analysis of consequences and estimation of significance when making decisions.

60 **communication** : successful transmission of information, data, energy (&tc.) along a message line, with a reception of feedback; an energetic flow of intention to cause an effect (or duplication) at a distance; the personal energy moved or acted upon by will or else 'selective directed attention'; the 'messenger action' used to transmit and receive energy across a medium.

61 **continuity** : being a continuous whole; a complete whole or "total round of"; the balance of the equation [− 120 + 120 = 0, &tc.]

own life, you practice the same for ALL Life at each Sphere of Existence, or else you ain't being truly Self-Honest.

This systematic process—as we present it in "Systemology"—begins with removal of emotional imprinting; all of which is coming from the Reactive Control Center (RCC) and so cannot be seen rationally[62] and analytically by the Master Control Center (MCC). This means that most people live their life in a reactionary fashion, often under the control of their emotions without even knowing it.

Here, it is important to mention, that what is implied by references to "emotions" really concerns the negative states of the Human Condition—which are all reactionary in nature—such as hopelessness, fear, anger, lust, jealousy, and so forth. For example: usually when you are angry, you are operating as a reaction to something—and the encoded mechanisms are commanded by the Reactive Control Center (plotted at "2.0" on the Standard Model or ZU-line). This is quite different from experiences of more positive states, such as being "in love" or personal enthusiasm about willingness[63] to act on something, &tc., which puts the individual at *cause*, rather than as the *effect*, and which are commanded by the Master Control Center (4.0).

Much of the Mardukite (NexGen) Systemology paradigm could be summed up as returning the Self to the state of being Cause rather than the Effect. You always should be able, at all times, by using the Master Control System through thought control, take notice when your Reactive Control Center takes over through emotions, and even correct accordingly with Self-Processing.* Eventually, pre-programmed automated reactivity is dissolved altogether—and that's Self-Honesty; *is* our "Systemology" in a nutshell. Again, there is much more to, but the important part to take away about these "levels" is that Self-Honesty is to be sought at each *gradient*—and the Seeker will quickly discover that these act as *Gateways* to accessing increasingly higher points of Actualized Awareness.[64]

One basic and effective practical method to systematically process "emotional imprinting" is described fully in "*The Tablets of Destiny*" (*Liber-One*) and outlined succinctly as "Systemology Process RR-SP-1." And there are many other practical processes within *Grade-III* material included in the follow up volume by Joshua Free: "*Crystal Clear*" (*Liber-2B*).

---

62 **rationality / reasoning (game theory)** : the extent to which a player seeks to play (make decisions, &tc.) in order to maximize the gains (or else survival) achievable within any given game conditions; the ability and willingness of an individual to reach toward conditions that promote the highest level of survival and existence and make the best choices and moves to see the desired goal manifest.

63 **willingness** : the ability and consideration to reach, face or confront some thing or energy; the ability and consideration to communicate along some line to produce an effect, to put attention or intention on the line.

* See *Liber-2B*, released as "*Crystal Clear*" by Joshua Free and contained in the "*Systemology Handbook.*"

64 **awareness** : the highest sense of-and-as Self in knowing and being as I-AM (the *Alpha-Spirit*); the extent of beingness directed as a POV experienced by Self as knowingness.

The basic theory supporting the Standard Model and *ZU-line* is a simple cosmology[65] rooted from the lore contained on *Arcane Tablets* from Mesopotamia.[66] Simply put: you have "AN"[67] which is the "*Spiritual*"; and "KI"[68] which is the "*Physical*"; between which exists a continuum[69] called "ZU,"[70] which manifests as "*Life*" or else "*Spiritual Life Awareness.*" A more specific discourse on the nature of "ZU" is the subject of a supplemental [delivered in lecture format by Joshua Free in December 2019 and then released in print form as "*The Power of Zu: Increasing Control of the Radiant Energy in Everyday Life*" with a foreword by Reed Penn.]

Great care has been taken with "*Tablets of Destiny*" (*Liber-One*) so that everything in the book is as clear a message as possible—even for a novice—including a concise definition of each word that could be problematic or misunderstood, and also definitions for the vocabulary introduced newly in our Systemology. Furthermore, *Liber-One* includes a summary of each lesson, given at the end of a chapter for optimal clarity. I also found that these summaries are great for just a quick second reading and review.

Yes, we are aware that some people will see "*Tablets of Destiny*" and "*Crystal Clear*" as just another mere "Self-Help" book series—which from a certain perspective this *is* a "Self-Help" series—but my take on this is that apparent these "Self-Help" books containing "deep esoteric occult wisdom" makes for a great change from all those books posing as "deep esoteric occult wisdom" and yet turn out to be mere Self-Help books that provide "no help."

As it's written: "Don't take anything from these books by faith. Apply these principles dir-ectly to your life..." to confirm that these are true for you from the perspective of Self and thus discover, and live, the life you were meant to live, Self-Honestly as a Free Spirit.

---

65 **cosmology** : a philosophy defining the origins and structure of the universe.

66 **Mesopotamia** : land between Tigris and Euphrates River; modern-day Iraq.

67 *AN* : an ancient cuneiform sign designating the *'spiritual zone'*; the *Spiritual Universe*—comprised of spiritual matter and spiritual energy; a direction of motion toward spiritual *Infinity*, away from or superior to the physical (*'KI'*); the spiritual condition of existence providing for our primary *Alpha* state as an individual *Identity* or *I-AM-Self* which interacts and experiences *Awareness* of a *beta* state in the *Physical Universe* (*'KI'*) as *Life*.

68 *KI* : an ancient cuneiform sign designating the *'physical zone'*; the *Physical Universe*—comprised of physical matter and physical energy in action across space and observed as time; a direction of motion toward material *Continuity*, away from or subordinate to the Spiritual (*'AN'*); the physical condition of existence providing for our *beta* state of *Awareness* experienced (and interacted with) as an individual *Lifeform* from our primary Alpha state of Identity or *I-AM-Self* in the *Spiritual Universe* (*'AN'*).

69 **continuum** : a continuous *whole*; observing all gradients on a *spectrum*; measuring quantitative variation with gradual transition on a spectrum without demonstrating discontinuity or separate parts.

70 *ZU* : the ancient cuneiform sign designating an archaic verb—"*to know,*" "*knowingness*" or "*awareness*"; the active energy/matter of the "Spiritual Universe" (AN) that is experienced as *Lifeforce* or *consciousness* for entities existing in the "Physical Universe" (KI); "*Spiritual Life Energy*"; the spiritual energy present in the WILL of the actualized *Alpha-Spirit* in the "Spiritual Universe" (AN), which imbues its *Awareness* into the Physical Universe (KI), animating/controlling *Life* for its experience of *beta-existence* along an *Identity-continuum* called a *ZU-line*.

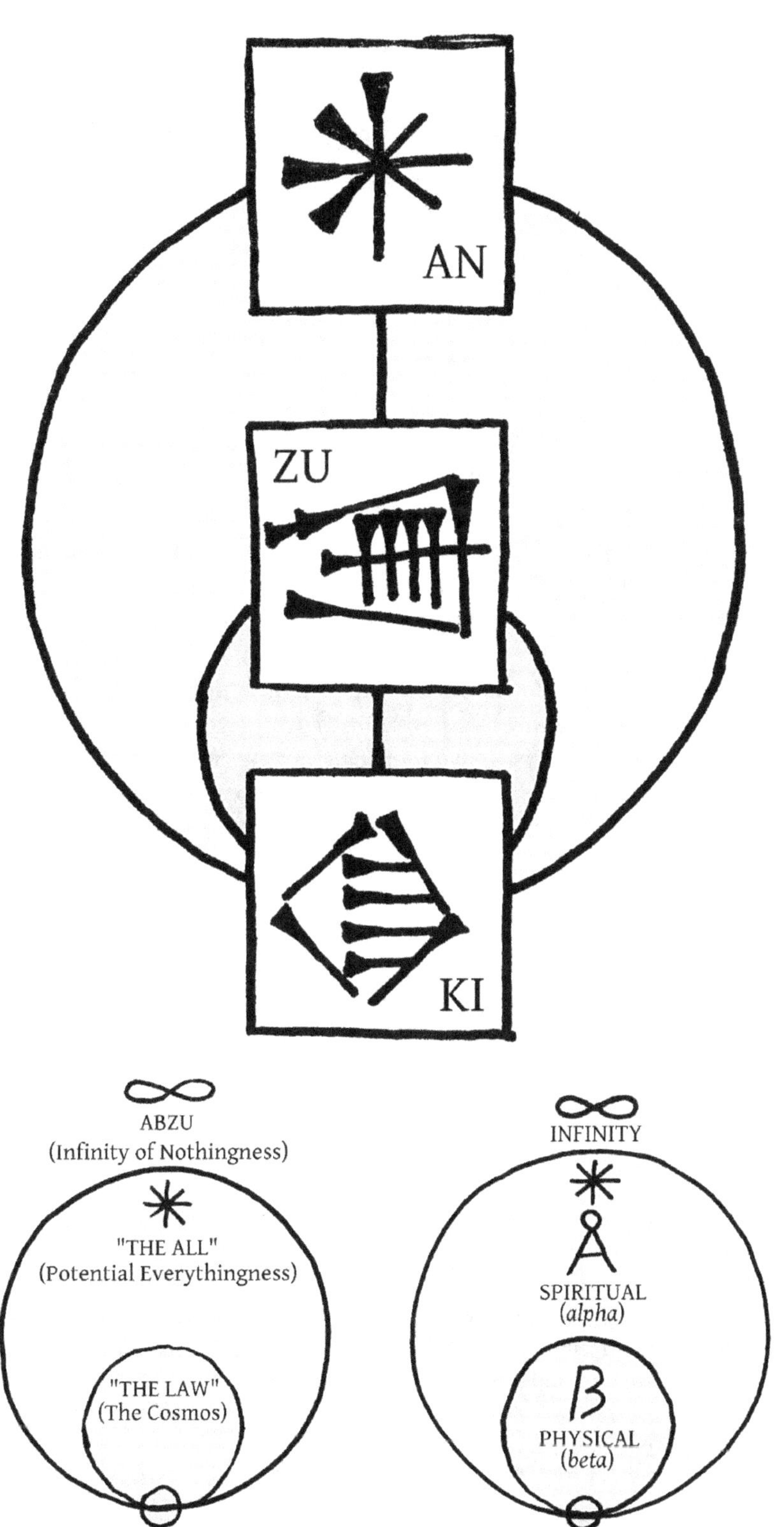

AN
ZU
KI
ABZU
(Infinity of Nothingness)
"THE ALL"
(Potential Everythingness)
"THE LAW"
(The Cosmos)
INFINITY
SPIRITUAL
(alpha)
PHYSICAL
(beta)

:: ✖ ::

MARDUKITE ZUISM & NEXGEN SYSTEMOLOGY
⟪ FREQUENTLY ASKED QUESTIONS ⟫
[Compiled & Answered by David Zibert*]

• *Why the "Necronomicon"?—Why do you use of the title "Necronomicon" to represent historical tablets for Mardukite work?*

In 1977, a mysterious book appeared in the underground occult scene, published by a collective related to the O.T.O.[71]—naming themselves "Schlangekraft"—which was edited with authorship under the single name: "Simon." That book, of course, being titled: "*The Necronomicon.*" Chances are good that you know of the book I'm talking about; the infamous "*Simon Necronomicon.*" Over a million print editions have circulated around the planet—not even including those now illegally copied and shared on the internet.

Behind, back of and beyond all of the controversy which surrounds the book, a Seeker will discover that a large part of its material is drawn from historical Mesopotamian cuneiform[72] tablets; for example: the *Enuma Eliš Babylonian Epic of Creation*, the *Fifty Names of Marduk* and the *Descent of Ishtar in the Underworld*, or *Sleep of Ishtar*—but all of which is bastardized to fit a stereotypical and expected "Lovecraftian" theme.

The "*Simon Necronomicon,*" up until the arrival of the Mardukite edition by Joshua Free in 2009, was actually an anomaly in the "New Age" scene: the sole book presenting a specifically Babylonian spiritual paradigm with any practical occult relevance to support a wide interest in "Mesopotamian Neopaganism." Joshua Free and the Mardukite Chamberlains team thus saw fit to give back the material its rightful due; providing us with the original Babylonian paradigm—called "Mardukite," as in dedication to the patron[73] Anunnaki founder of Babylon, Marduk.

---

* Based on a transcript of a video presentation delivered to the "Mardukite Chamberlains" and "NexGen Systemology Society" by David Zibert in November 2019.

71 **"Ordo Templi Orientis" (OTO)** : "Order of the Eastern Temple" or "Order of the Eastern Templars"; a underground occult German-inspired variation on masonic-styled "Golden Dawn and Rosicrucian type" methodology; founded by Theodor Reuss (*et al.*) in *c.*1900, and within a decade included such membership as Gerald Gardner, Aleister Crowley and other early prominent "New Age" figures; surviving mainstream OTO factions are those strongly influenced by Crowley's "Thelemic" work and later "Typhonian" contributions by his heir-apparent, Kenneth Grant.

72 **cuneiform** : the oldest extant writing from Mesopotamia; wedge-shaped script inscribed on clay tablets with a reed pen.

73 **patron god** : the most sacred deity of a region or city, of which most temples and religious services are directed; the personal deity of an individual.

This new wave of "Mardukite Necronomicon" work by Joshua Free was firmly based on actual cuneiform clay tablet collections, complete with initiate-level commentary available to the public for the first time; replacing the need to independently and meticulously collect dry and often hard to find academic renditions of similar material. The original installment of this work was released on the Summer Solstice 2009 exclusively to the "Mardukite Chamberlains" as *"Liber-N: The Necronomicon of Joshua Free"*—which is actually not a derivative of "Simon's" edition and really has nothing directly to do with the stories of H.P. Lovecraft. The official Mardukite comment on "Simon's" book was published in 2012 as *Liber-555*, supplementing the 2010 release of *Liber-R*.[‡]

• *Is Mardukite Zuism an Ancient Aliens Cult?*

No, we aren't—We also do not specifically or exclusively subscribe to Zecharia Sitchin's interpretation of the Anunnaki, either. We *do* view the historic Anunnaki as physical beings, *yes*, but beings maintaining an elite quality of Actualized Awareness as "Alpha Spirits" in command of a "genetic vehicle" or physical body, instead of literal "spacemen"—which they just might be too, as everything in the Physical Universe carries "multidimensional" facets, at least from the perspective of (and understanding held by) the Human Condition.

While it is true that Mardukite Zuism is indeed a recreation, reconstruction or reenactment of the Ancient Babylonian Priesthood of Nabu, we are not worshiping the Anunnaki as divinity, nor in the conventional sense, Marduk directly proper. Relevance of the *"Babili"* System—the native name for the Babylonian spiritual paradigm—lies in it's underlying system and it's logic—or "Systemology"—that we are now applying in modern times. It would be useless and unproductive to merely reiterate and celebrate the ancient tradition verbatim, which purpose is now obsolete.

It could further be said that the way in which we process this material now in the 21st century AD—the stuff coming to us from the 21st century BC—is a practice of "defragmentation"; that we are actually very much systematically undoing the fragmentation once done back in Babylonia, in some kind of esoterically revolutionary 'Averse Mardukite' approach, a semantic we are of course avoiding because it's misleading.

In brief, "Mardukite Zuism" and its *Systemology* is not a cult; quite the opposite—it is a map by which we reclaim the highest potentials of the Self—the Alpha-Spirit—or, as we say in a rather poetic fashion: a way to "Starwalk to the Source."

[‡] "Liber-555" and "Liber-R" have since been combined for the most recent 2019 edition of *"Necronomicon Revelations"* by Joshua Free; also available in the *"Gates of the Necronomicon"* (2019 Hardcover) anthology by Joshua Free and the 2019 Grade-II Master Edition release of *"Necronomicon: The Complete Anunnaki Legacy."*

• How do you "join" the Mardukites?

Simply put, in the most basic terms: you don't. You *read* the books; you *do* the work; you stick around and if you get it right—as in what I call the *'Mardukite Key'*—then yeah, congratulations, you're a "Mardukite." This is not to say that there won't be more structure within the organization in the future—such as we see progressing with our "Grades" and the development of the "Mardukite Academy," "International School of Systemology" and "Mardukite Zuism" as an openly practiced modern spiritual paradigm—but we've found that, for the most part, a classical "masonic" or "kabbalistic" structure of initiation, such as is widely seen in Western culture, is more often a distraction, a game[74] really, rather than the real deal, which is an association of individuals maintaining "Actualized Awareness." That being said, we are expecting and preparing for increased organizational involvement throughout the 2020's decade and beyond!

• How many freakin' books does this Joshua guy put out every year?

This is a more commonly asked question by those individuals that have only recently, within the past couple of years, become aware of the existence of Mardukite Ministries, Mardukite Systemology and the work of Joshua Free as newly republished and globally distributed through the new "Joshua Free Publishing Imprint." Newcomers can be easily overwhelmed by the sheer amount of materials released over the short period of time that this new imprint has been in operations; more precisely, at least 20 ISBN assigned in about 15 months, which is indeed quite more than other esoteric and occult books publishers that had started up around the same time (Samhain 2018).[∞]

What some might be unaware of, is that most of the books released between 2018 and now are actually rewritten, revised and reissued materials that were originally published over the past decade from "Mardukite Truth Seeker Press" (founded in 2008), which was the original underground print-on-demand version of the indie publishing house now represented by the official Joshua Free Imprint. Hence you now see these vastly improved versions as more commercially available "10th anniversary" or "20th anniversary" collector's editions—many of which now appear in hardcover for the very first time. The *Grade-II* "Mardukite Core" *is* composed of a decade of material now reissued back to back, so I do understand why some newcomers might find it overwhelming. Even then, if you look back at the last decade of Mardukite and Systemology work, Joshua Free has always been releasing stuff at a frantic pace, something largely unseen on the scene in general since the end of the 20th century.

---

74 **game** : a strategic situation where the power of choice is employed or affected; a parameter or condition defined by purposes, freedoms and barriers (rules).

∞ The "Joshua Free Imprint" was founded with the release of its first official title—*"Necronomicon: The Anunnaki Bible"* (Hardcover 10th Anniversary Collector's Edition)—on October 31, 2018.

• *Are the Mardukites RHP (Right-Hand Path) or LHP (Left-Hand Path)?*

We're neither. While it is true, we often describe our path—which is the "*Pathway to Self-Honesty*"—as the "Right Way," this semantic is often confused with an RHP context; which it is not. These polarized terms of "Left" and "Right" have actually been obsolete for quite a while in the underground—and I expect them to also become so on the more mainstream scene somewhat soon, as there are signs of that here and there already. In brief—"what is termed LHP is really a blend of the diabolized traditions of Enki and Ishtar, but realized very much outside of Self-Honesty, while the RHP is a corrupted expression of an usurpation of the Mardukite Ideal, which in itself predates all this."[*] Do your homework and you'll see that this is very much the case.

• *What are the sources of all these Arcane Tablets?*

I get asked this a lot from folks who would like a more academic presentation of the material that appears in our "*Necronomicon: Anunnaki Bible*" or "*The Complete Anunnaki Bible*" (Etc.)—people that obviously are unaware of how things work in the occult underground. But *yes!*—all of these tablets and their contents may be independently researched from the remaining physical cuneiform clay tablet collections from ancient Mesopotamia, all of which are kept in museums; with the exception, perhaps, of the "Shayaha" (Sajaha) materials,[‡] the source of which is still the subject of some controversy.

The more widely accessible *Arcane Tablets* found within the "Mardukite Catalogue" or text of "*The Complete Anunnaki Bible*" are based on updated translations of transliterations from the "Kuyunjik" collection and "Surpu" tablet series kept in the British Museum, in addition to the "Maqlu" and other now well-known sources of cuneiform literature. It is all easily researched and readily discoverable if you actually care to do so; but officially, we have already put in the time on these subjects. I might add that: the very fact that the core of our work is based on real archaeology—on very raw physical data—and not some Illuminoid mumbo jumbo, demonstrates both a seriousness and validity[75] of what we're doing, regardless of your opinion of, for example, our use of the title "*Necronomicon*."

• *What is the link between Babylonia and Egypt?*

---

[*]   Paraphrasing from the essay "*The Game of True Knowledge*" by David Zibert published in the *Pillars* occult periodical by Anathema Publishing.

[‡]   "*Mardukite Tablet-S Series*" is found in the *Grade-II* Mardukite Core and also in the appendix of "*The Tablets of Destiny*" (*Mardukite Systemology Liber-One*). It was allegedly unearthed during German excavations of Babylon and then came into the possession of the "Vril Society."

75 **validation** : the reinforcement of agreements of Reality.

If you read *"The Complete Anunnaki Bible,"* you'll notice a considerable amount of material comes from Egyptian sources to supplement the primary Mesopotamian emphasis of the work—and people often wonder how the two are related. Well, it must first be understood that what we term "Mardukite" relates to a very specific line of Mesopotamian kings across a very specific timeline. Some of the most well known Mardukite Babylonian kings are, of course, Sargon, Nebuchadnezzar II and Ashurbanipal. Their time of reign are best characterized by material and spiritual prosperity of their lands, as opposed to a focus on brutal warfare and conquest. This is reflected in the cuneiform records, such as is found on the famous *"Nebuchadnezzar II Tablet"* as cataloged in the Mardukite Tablet-L Series:[∞]

> Lord *Marduk*, Glorious Chief of the *Anunnaki* has heard my petition.
> I have subdued those who do not heed to the Will[76] of *Marduk*.
> I have overcome the rebels in the country, and made the inhabitants of my kingdom
> Loyal through their prosperity.
> To *Babylon*, I have carried gold, silver, copper, lapis lazuli
> And other precious stones and expensive woods [resins] from the mountains.
> I have decorated the Shrine of *Marduk*, The *Esagila*,
> The Shrine of *Marduk*, I have decorated with gold.
> Wood from the finest cedars of *Lebanon* comprise the roof.
> And I have decorated the "Boat" of *Marduk*.
> The Gate of *Nanna-Sin* [the Moon god], I have plated with Silver.
> The AIBURSHABU [Processional Way of *Marduk*], I have paved with tiles.
> Since *Marduk* created me to be king,
> And *Nabu* has culled his people to my realm,
> As the love I have for my own life,
> So do I feel toward the building and reign of their cities.
> Lord *Marduk*, Captain among the *Anunnaki* has received my prayer.

And from the same Mardukite Tablet-L Series collection, we read *"The Sargon Tablet"* scribed over a millennium before the time of Nebuchadnezzar II:

> I am *Sargon*, the Mighty King of Akkad.
> My mother was of humble estate, and I knew nothing of my father.
> The brother of my father was a Dweller in the Mountain.
> My city is *Azur-Pirani* (Which lies on the banks of the Euphrates)
> My humble mother brought me forth in secret.
> In a basket of reeds she laid me, she smeared the latch with bitumen
> And pushed me off into the river.

∞ *"The Complete Anunnaki Bible"* edited by Joshua Free.

76 **will** *or* **WILL** (5.0) : in *NexGen Systemology* (from the *Standard Model*)—the spiritual ability at (5.0) of an *Alpha Spirit* (7.0) to apply *intention* as "Cause" from a higher order of reasoning and consideration (6.0) than the thoughts found in *beta-existence,* where it manifests as "effect" below (4.0).

But the river did not swallow me.
The river carried me to *Akki* (A man who labored by watering the fields)
Akki lifted me out of the basket, raised me as his son.
Akki reared me to be his gardener,
And all the while I gardened, *Marduk* smiled on me.
And by his love, I was made a Ruler of the Kingdom.

So where did the trouble begin? Well, what happened is that Marduk's systematization in Babylonia was not exactly well received by other members of the younger Anunnaki Pantheon; so basically he got kicked out and imprisoned—the official reason being that he went against the established *"World Order of Enlil"* by marrying a human offspring instead of his half-sister Ishtar. This is, again, all found on the *Arcane Tablets* within the Mardukite catalogue; specifically the *"Erra Epos"* series found as *"Mardukite Tablet V"* in *"The Complete Anunnaki Bible."*

So what Marduk did—with the help of Nabu, his heir-son—was to temporarily move his legacy to Egypt through, once again, a very specific dynasty where he was first known as "Amon-RA," then as the sun god "Aten"—the flavor of spiritual and political reign of Akhenaten being one-to-one[77] with those of the aforementioned Babylonian kings. While the rule of these Egyptian Mardukites was also overthrown within a short time, their famous legacy is not to be overlooked, as it contains specific information which remained absent from it's Babylonian counterpart.[*]

• *Is there a link between the Anunnaki paradigm and (insert preferred cultural mythology here)?*

Perhaps one of the most commonly proposed questions by those who have studied various mythologies and cultures is: are the Anunnaki related to India, or Japan, or Mesoamerica, &tc. The short answer is: *yes*—the Babylonian Anunnaki are planetary archetypes,[78] the same as those found throughout each and every culture worldwide, without exception. And while this is easily demonstrated intellectually with small investigations into comparative religion, it often comes as an emotional shock to those that fixedly identify themselves very strongly to a specific localized[79] culture (and its correlated[80] mythology semantics). These are the same individuals that would dismiss ancient Mesopotamian lore today because of the current politics in the Middle-East and its present population.

---

77 **A-for-A (one-to-one)** : meaning that what we say, write, represent, think or symbolize is a direct and perfect reflection of the actual aspect or thing—that "A" is for, means and is equivalent to "A" and not "a" or "q" or "!"

* See Mardukite Liber-V, *"The Vampyre's Handbook"* by Joshua Free.

78 **archetype** : a "first form" or ideal conceptual model of some aspect; the ultimate prototype of a form on which all other conceptions are based.

79 **localized** : brought together and confined to a particular place.

80 **correlating** : the relationship between two or more aspects, parts or systems.

Just do your homework in Self-Honesty and the facts will become clear.

But, to be a bit more specific concerning the worldwide influence of the Anunnaki, there are a few examples given in the Mardukite literature concerning this if you look. One such is Ningishzidda, which is Enki's brother, that becomes a driving force behind the inception of ancient Meso-American civilization under the guise of Quetzacoatl. Other details can be found regarding the Tuatha d'Anu in Europe that also carried the Anunnaki Legacy in the guise of "Druidism."[‡] Essentially, one can find connections between anything and anything. But, such connections and cataloging is not the focus of our current work, as stated by Joshua Free in the original Mardukite Systemology *Grade-III* textbook, *Liber-One:*[∞]

> "Once an individual's attention[81] is successfully brought to the wisdom of the *Arcane Tablets*, suddenly new layers of understanding begin to unfold concerning Life, the Physical Universe—and experience of it as Reality. With new eyes, the Seeker begins to find fragments of confirmation in virtually every piece of ancient esoterica. This information is not restricted only to the cuneiform tablets unearthed in Mesopotamia —but it is our oldest, complete and definitive literary example, from which sprung the "*Secret Doctrines*" maintained by many other spiritual traditions and cultures throughout time[82] and space.

> "[83]Of course, if we choose to, we will undoubtedly find variations of this wisdom in the *Egyptian Book of the Dead*, the *Dhammapada*, the *Vedas*, and so on… With humanity now having wide access to various source materials, it is not the focus or intent of our current discourse—or the instructional Systemology Grade that it represents—to chase down all later derived fragmented examples of the "*Tablets of Destiny*" in history. Such is an intellectual occupation that someone else will undoubtedly take up, but no attempts will be made at this time by the current author to systematically redefine and categorize all of the diverse paradigms in existence."

---

‡   See the Mardukite Liber-D Trilogy by Joshua Free, specifically "*Elvenomicon*" and "*Draconomicon*" both of which are also found in the Grade-I "Route of Druidism" Master Edition release of "*Merlyn's Complete Book of Druidism: A Master Course in Druidry for Modern Druids.*"

∞   Available as "*The Tablets of Destiny: Using Ancient Wisdom to Unlock Human Potential*" and included in the complete Grade-III Master Edition release of "*The Systemology Handbook*" by Joshua Free.

81 **attention** : active use of *Awareness* toward a specific aspect or thing; the act of "attending" with the presence of *Self*; a direction of focus or concentration of *Awareness* along a particular channel or conduit or toward a particular terminal node or communication termination point; the Self-directed concentration of personal energy as a combination of observation, thought-waves and consideration.

82 **time** : observation of cycles in action; motion of a particle, energy or wave across space; intervals of action related to other intervals of action as observed in Awareness; a measurable wave-length or frequency in comparison to a static state; the consideration of variations in space.

83 **space** : the viewpoint (or POV) extended out from any point out toward a dimension or dimensions; the consideration of a point or spot; the field of energy/matter mass created as a result of communication and control in action and measured as time (wave-length).

• *How is Mardukite Zuism & Systemology any different from other esoteric occult, initiatory or religious organizations?*

This is certainly a legitimate and relevant question. The Mardukite material, in itself, reveals what I like to call the *'Mardukite Key'*—as I mentioned above—and it is this very *Key* that makes our work and organization stand apart from anything that has publicly come before. This whole matter is somewhat complex—a bit hard to define fully in only a few words. Simply put, the *'Mardukite Key'* attained from *Grade-II* work is what we call Systemology, which is explored directly in the *Grade-III* work, including "*The Tablets of Destiny*" and "*Crystal Clear.*"

In effect, what we have found with the recension and study of the most ancient texts available is how, primarily through the very use of the written word, the Human Condition got programmed outside of a Self-Honest experience of reality.

What do we mean by Self-Honesty? The term is clearly defined in the "Mardukite Zuism & NexGen Systemology Dictionary"—

"the *alpha*[84] state of *being* and *knowing*; clear and present total *Awareness*[85] of-and-as *Self*, in its most basic and true proactive expression of itself as *Spirit* or *I-AM*—free of artificial attachments, perceptive filters and other emotionally-reactive or mentally-conditioned programming imposed on the Human Condition by the systematized physical world."

Whether or not this—what might be called the "*Babylon Incident*"—is literally the first time such programming was impressed on humanity, is actually irrelevant. It is historically appropriate for our purposes to chart the progression of implanted fragmentation programs, described in our Systemology, as they *were* visibly installed and are still running today, very much present in the Human Condition and its societies.

The exact nature, structure and control of these "systems" is often debated, understandably so, even in ancient times, when several members of the younger Anunnaki pantheon attempted the same tactics as Marduk. This often occurred simultaneously and using the same technology, known from the *Arcane Tablets* as the "Divine *ME*," or else the "*Tablets of Destiny.*"* Aside from Marduk's Babylon, we find similar patterns among the Anunnaki with

---

84 **alpha** : the first, primary, basic, superior or beginning of some form; in NexGen systemology, it also refers
    to the state of existence that operates on archetypes, will and intention "exterior" to the low-level
    condensation and solidarity of energy and matter as the 'physical universe'.

85 **awareness** : the highest sense of-and-as Self in knowing and being as I-AM (the *Alpha-Spirit*); the extent of
    beingness directed as a POV experienced by Self as knowingness.

* See "Mardukite Systemology Liber-One" available as "*The Tablets of Destiny: Using Ancient Wisdom to
    Unlock Human Potential*" and included in the complete *Grade-III* Master Edition release of "*The
    Systemology Handbook*" by Joshua Free.

the cases of Ishtar and Ninurta.

So, 'why the emphasis on Marduk then'—you may as well ask?

The simplest and most direct answer: because he is the only one among the Anunnaki that seemed to care at all about humanity; seeing the potential of the species as more than mere cattle or pets in material service to the gods—and their plans. As mentioned within the cycle of Mardukite literature, or the Mardukite Core,‡ *his own plan*—to return all religious devotion of Humanity to the Infinite Source, or else activating complete personal *Alpha Awareness* of the individual through all Spheres of Existence—*failed.*

And this is precisely what we, the modern Mardukite movement, have set up to correct right now, using the applied spiritual technology that we call "Systemology"—the very same technology used by the Anunnaki to engineer the realities of *this* Physical Universe. Furthermore, as I mentioned in my foreword to the French translation of "*Mardukite Liber-G*"—the ancient Mardukite Babylonian paradigm (and specifically its Systemology) from the 21st century BC reveals nothing short of a more complete, workable, effective and valid proto-kabbalah, predating the better known and widely used Semitic version by hundreds, if not thousands, of years.

Full realization of the '*Mardukite Key*' is not to be taken lightly, as it positively cancels centuries, even millennia of fragmented exegesis. This simple fact, dear Seekers and friends, is a major game-changer for humanity—at the very least, as far as esoteric literature and the occult underground is concerned; something that has yet to be fully realized for essentially political reasons. And as I said in my contribution to "*Mardukite Liber-51*"∞ many years ago: this time, it's different, and we are cheerfully inviting you to join us and take part of this grand adventure, this real hope for the future, to reclaim your True Self, and evolve, and rise, as *Homo Novus*,[86] or else go extinct alongside *Homo-Sapiens* and it's pre-programmed obsolescence.

---

‡ The *Grade-II* "Mardukite Core" or "Necronomicon Cycle" is collected (15 books) in one volume as: "*Necronomicon: The Complete Anunnaki Legacy*" (2020 Master Edition) by Joshua Free.

∞ "*Liber-51*" is contained in the 2019 paperback edition of "*Sumerian Legacy*" by Joshua Free, the 2019 hardcover edition of "*Gates of the Necronomicon*" and the complete *Grade-II* Master Edition hardcover "*Necronomicon: The Complete Anunnaki Legacy.*"

86 **Homo Novus** : literally, the "new man"; the "newly elevated man" or "known man" in ancient Rome; the man who "knows (only) through himself"; in NexGen Systemology—the next spiritual and intellectual evolution of *homo sapiens* (the "modern Human Condition"), which is signified by a demonstration of higher faculties of *Self-Actualization* and clear *Awareness*.

NOW YOU KNOW!

>> LIBER 3C <<

BY JOSHUA FREE

:: 1 ::

APPROACHING THE GATES OF UNDERSTANDING
« UNIFYING MARDUKITE ZUISM & SYSTEMOLGY »

Many esoteric[87] models of universal cosmology[88] and spiritual ascension[89] appear on the timeline[90] of acutely recorded history over the past 6,000 years, and the most ancient of these records—the *Arcane Tablets*—reveal the most simple and complete account of Cosmic History,[91] that which later inspired an entire planet of cultural mythologies and religious interpretations. Yet none of these further fragments[92] and facets[93] of the original *Crystal* ever brought a clearer experience or more perfect understanding[94] than what had come before. As a result, the truth inherent in the simplicity once shared became forgotten and lost to a sea of "symbols" and "representations" that reflected the poorest shadows of a former age—and the *Ancient Mystery School*[95] was born.

---

87 **esoteric** : hidden; secret; knowledge understood by a select few.

88 **cosmology** : a philosophy defining the origins and structure of the universe.

89 **ascension** : actualized *Awareness* elevated to the point of true "spiritual existence" exterior to *beta existence*. An "Ascended Master" is one who has returned to an incarnation on Earth as an inherently *Enlightened One*, demonstrable in their actions—they have the ability to *Self-direct* the "Spirit" as *Self*, just as we are treating the "Mind" and "Body" at this current grade of instruction; in *Moroii ad Vitam*, a state of Beingness after *First Death*, experienced by an *etheric body*, which is able to maintain consciousness as a personal identity continuum with the same *Self-directed* control and communication of Will-Intention that is exercised, actualized and developed deliberately during one's present incarnation.

90 **timeline** : plotting out history in a linear (line) model to indicate instances (experiences) or demonstrate changes in state (space) as measured over time; a singular conception of continuation of observed time as marked by event-intervals and changes in energy and matter across space.

91 **Cosmic History** : the entire timeline of all existence, starting with the Infinity of Nothingness and individuation of Self and ending with the condensation and solidification of this Physical Universe experienced in present-time.

92 **fragmentation** : breaking into parts and scattering the pieces; the *fractioning* of wholeness or the *fracture* of a holistic interconnected *alpha* state, favoring observational *Awareness* of perceived connectivity between parts; *discontinuity*; separation of a totality into parts; in *NexGen Systemology*, a person outside a state of *Self-Honesty* is said to be *fragmented*.

93 **facet** : an aspect, an apparent phase; one of many faces of something; a cut surface on a gem or crystal; in *NexGen Systemology*—a single perception or aspect of a memory or "*Imprint*"; any one of many ways in which a memory is recorded; perceptions associated with a painful emotional (sensation) experience and "*imprinted*" onto a metaphoric lens through which to view future similar experiences; other secondary terminals that are associated with a particular terminal, painful event or experience of loss, and which may exhibit the same encoded significance as the activating event.

94 **understanding** : a clear 'A-for-A' duplication of a communication as 'knowledge', which may be comprehended and retained with its significance assigned in relation to other 'knowledge' treated as a 'significant understanding'; the "grade" or "level" that a knowledge base is collected and the manner in which the data is organized and evaluated.

95 **Ancient Mystery School** : the original arcane source of all esoteric knowledge on Earth, concentrated between the Middle East and modern-day Turkey and Transylvania c. 6000 B.C. and then dispersing south (Mesopotamia), west (Europe) and east (Asia) from that location.

Patterns demonstrated by this marked descent of civilization are cyclic in nature and apply to all "systems"—including the *condensation*[96] of "universes." Some individuals[97] classify tendencies of this "downward spiral" using semantics[98] for energy, such as "entropy"[99]—whereas others think in terms of material "degradation." Regarding a relative[100] direction between states or conditions,[101] there are also those that refer to these motions as "condensation" and "evaporation." In our Systemology, we use a "Standard Model"[102] (also treated as the "ZU-line"[103] when it applies to the individual "*Self*") with a systematic continuum[104] between "zero" and "Infinity." We use this to easily demonstrate understanding of varying gradients:[105] conditions of existence[106] and "degrees"[107] by which they are experienced.

---

96 **condense / condensation** : the transition of vapor to liquid; a change in state demoting a more substantial or solid condition; leading to a more compact or solid form.

97 **individual** : a person, lifeform or human entity; a *Seeker* or potential *Seeker* is often referred to as an "individual" within Mardukite Zuism and Systemology materials.

98 **semantics** : the *meaning* carried in *language* as the *truth* of a "thing" represented, *A-for-A*; the *effect* of language on *thought* activity in the Mind and physical behavior; language as *symbols* used to represent a concept, "thing" or "solid."

99 **entropy** : the reduction of organized physical systems back into chaos-continuity when their integrity is measured against space over time.

100 **relative** : an apparent point, state or condition treated as distinct from others.

101 **condition** : an apparent state; circumstances and dynamics that affect the order and function of a system; a series of interconnected requirements, barriers and allowances that must be met; in traditional language, to bring a thing toward a specific, desired or intentional new state (such as in conditioning), though to minimize confusion, *NexGen Systemology* usually treats this with the semantics of imprinting, encoding and programming.

102 **standard model** : a fundamental *structure* or symbolic construct used to evaluate a complete *set* in *continuity* relative to itself and variable to all other *dynamic systems* as graphed or calculated by *logic*; in *NexGen Systemology*—our existential and cosmological cabbalistic model; a "*monistic continuity model*" demonstrating *total system* interconnectivity "above" and "below" observation of any apparent *parameters*; the *ZU-line* represented as a singular vertical (*y*-axis) waveform in space across dimensional levels (universes) without charting any specific movement across a dimensional time-graph *x*-axis.

103 ***Zu*-line** : a spectrum of *Spiritual Life Energy (ZU)* as conceived on the Standard Model of Systemology; an energetic channel of *Identity-continuum* connecting the *Awareness (ZU)* of an *Alpha-Spirit* with "*Infinity*"; a *Life-line* on which *Awareness (ZU)* extends from the direction of the "Spiritual Universe" (AN) as its *alpha state* through an entire possible range of activity in its *beta state*, experienced as a *genetic-entity* occupying the *Physical Universe (KI)*; the Standard Model demonstrates the Zu-line interacting with spheres of existence.

104 **continuum** : a continuous *whole*; observing all gradients on a *spectrum*; measuring quantitative variation with gradual transition on a spectrum without demonstrating discontinuity or separate parts.

105 **gradient** : a degree of partitioned ascent or descent along some scale, elevation or incline; "higher" and "lower" values in relation to one another.

106 **existence** : the *state* or fact of *apparent manifestation*; the resulting combination of the Principles of Manifestation: consciousness, motion and substance; continued *survival*; that which independently exists; the 'Prime Directive' and sole purpose of all manifestation or Reality; the highest common intended motivation driving any "*Thing*" or *Life*.

107 **degree** : a physical or conceptual *unit* (or point) defining the variation present relative to a *scale* above and below it; any stage or extent to which something *is* in relation to other possible positions within a *set* of "*parameters*"; a point within a specific range or spectrum; in *NexGen Systemology*, a *Seeker's* potential

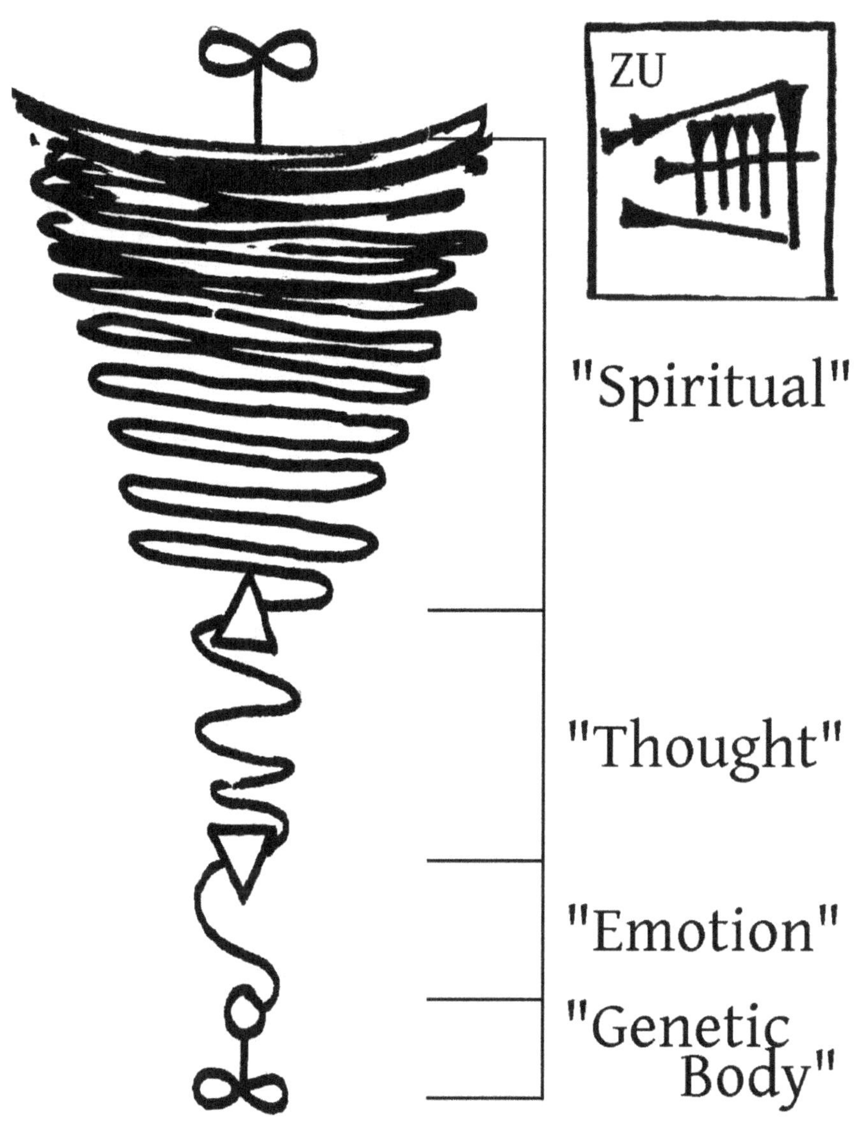

____

energy variations or fluctuations in thought, emotional reaction and physical perception are all treated as
"degrees."

As we move our considerations[108] or "POV"[109] (point-of-view or viewpoint) <u>closer</u> to "zero"—which generally represents the basic physical continuity[110] of beta-existence, or the *stuff* of this "Physical Universe"—the space-time energy-matter is more fragmented, condensed and fixed as a solid. When we apply the same model to the individual's "beta" *Awareness* level, we say that they are "withdrawing" attention[111] and *Awareness*[112] and thus becoming more of the "effect" of the Physical Universe. Similarly, as we consider points <u>further</u> away from "zero," the same energetic patterns[113] appear more "vapor-like" and "fluid" and increase in their "potentiality" as an existence or potential beingness as a POV. Likewise, an individual's *Awareness* increases along with their "reach" in the direction of "cause," which is to say true *Actualization*[114] and *Self-determinism*[115] and to the extent they may Self-Honestly project[116] their POV.

Although considerable discrepancy in semantics, vocabulary and human understanding exists in regards to our Cosmic History and the original *map* and *key* left to us on the *Arcane Tablets*, most of this "mythology" and "symbolism" has previously only been used as a basis of lesser purposes—including the further solidification and fragmentation of an individual's considerations as they occupy this beta-existence.

---

108 **consideration** : careful analytical reflection of all aspects; deliberation; determining the significance of a "thing" in relation to similarity or dissimilarity to other "things"; evaluation of facts and importance of certain facts; thorough examination of all aspects related to, or important for, making a decision; the analysis of consequences and estimation of significance when making decisions.

109 **point-of-view (POV)** : an opinion or attitude as expressed from a specific identity-phase; a specific standpoint or vantage-point; a definitive manner of consideration specific to an individual phase or identity; a place or position affording a specific view or vantage; circumstances and programming of an individual that is conducive to a particular response, consideration or belief-set (paradigm); a position (consideration) or place (location) that provides a specific view or perspective (subjective) on experience (of the objective).

110 **continuity** : being a continuous whole; a complete whole or "total round of"; the balance of the equation [$-120 + 120 = 0$, &tc.]

111 **attention** : active use of *Awareness* toward a specific aspect or thing; the act of "attending" with the presence of *Self*; a direction of focus or concentration of *Awareness* along a particular channel or conduit or toward a particular terminal node or communication termination point; the Self-directed concentration of personal energy as a combination of observation, thought-waves and consideration.

112 **awareness** : the highest sense of-and-as Self in knowing and being as I-AM (the *Alpha-Spirit*); the extent of beingness directed as a POV experienced by Self as knowingness.

113 **patterns (probability patterns)** : observation of cycles and tendencies to predict a causal relationship or determine the actual condition or flow of dynamic energy using a holistic systemology to understand Life, Reality and Existence as opposed to isolating or excluding perceived parts as being mutually separate from other perceived parts.

114 **Self-actualization** : bringing the full potential of the Human spirit into Reality; expressing full capabilities and creativeness of the *Alpha-Spirit*.

115 **Self-determinism** : the freedom to act, clear of external control or influence; the personal control of Will to direct intention.

116 **projecting awareness** : sending out (motion) or radiating "*consciousness*" from *Self* ("I") to another POV.

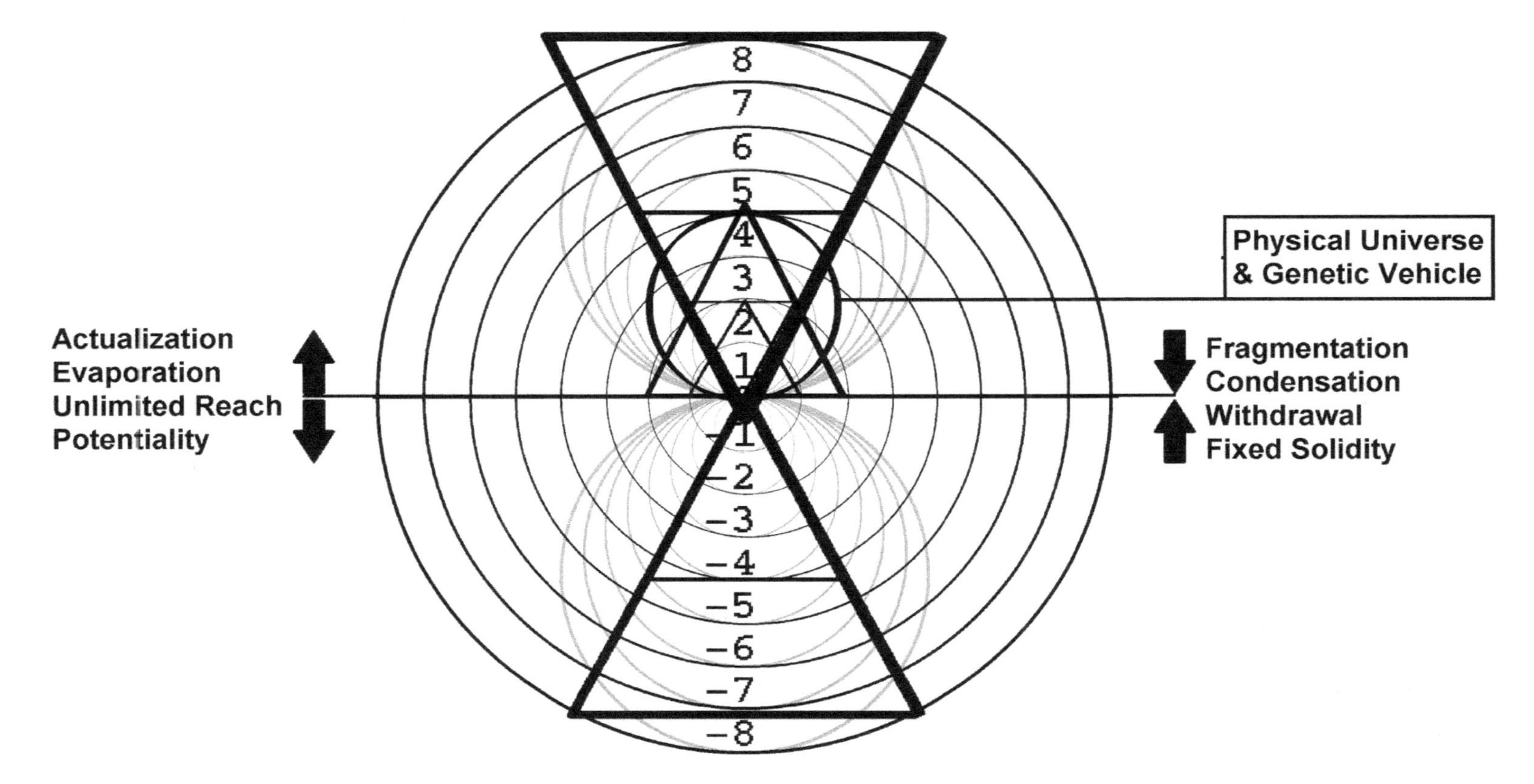
Physical Universe
& Genetic Vehicle
Fragmentation
Condensation
Withdrawal
Fixed Solidity
Actualization
Evaporation
Unlimited Reach
Potentiality
8
7
6
5
4
3
2
1
-1
-2
-3
-4
-5
-6
-7
-8

This is one of the primary concerns with basing our Systemology on the Standard Model or any fixed paradigm:[117] that its classification and demonstration of "divisions," "levels"[118] and "layers" of existence will be over-identified as "symbols" that poorly substitute the true understanding as a *knowing*.

Foundations of our higher graded work—including this present *Systemology Grade-IV*—is grounded firmly on the basis of research and discoveries presented in *Grade-III*, particularly as introduced in "*The Tablets of Destiny*" (*Liber-One*) and its companion manual "*Crystal Clear*" (*Liber-2B*).‡ The entire premise[119] of this "Standard Model" and "ZU-line" is established within those texts, supplemented by suggestions for practical systematic "processing" that supports each installment of instruction. The subject of the "processing" itself, and the complete course on "Communication, Control and Command" is what opens our present *Systemology Grade-IV*, with the previous publications: "*Communication and Control of Energy & Power*" (*Liber-2C*) and "*Command of the Mind–Body Connection*" (*Liber-2D*).

The most ancient recorded chronicle of our Cosmic History on Earth—that which includes cosmological information predating even the existence of *Life* on Earth—is best found on cuneiform[120] tablets; and among these, the "*Babylonian Epic of Creation*" known to scholars as the *Enuma Eliš*, so named for its opening lines. Unfortunately, even in ancient Babylon, these and other *Arcane Tablets* functionally assisted those that sought further fragmentation and successive[121] programming of the Human Condition[122] rather than liberating it. This is another pattern that we have seen many times since whenever similarly derived paradigms sought to provide any aid to the spiritual "Rescue Mission" (via "defragmentation[123] of the Human Condition") presently taking place in the Physical Universe and on Earth.

---

117 **paradigm** : an all-encompassing *standard* by which to view the world and *communicate* Reality; a standard model of reality-systems used by the Mind to filter, organize and interpret experience of Reality.

118 **level** : a physical or conceptual *tier* (or plane) relative to a *scale* above and below it; a significant *gradient* observable as a *foundation* (or surface) built upon and subsequent to other levels of a totality or whole; a *set* of "*parameters*" with respect to other such *sets* along a *continuum*; in *NexGen Systemology*, a Seeker's understanding, *Awareness* as *Self* and the formal grades of material/instruction are all treated as "*levels*."

‡ Materials from both texts are collected in the complete Grade-III anthology, "*The Systemology Handbook*" by Joshua Free (with additional contributions by Reed Penn and Kyra Kaos).

119 **premise** : a basis or statement of fact from which conclusions are drawn.

120 **cuneiform** : the oldest extant writing from Mesopotamia; wedge-shaped script inscribed on clay tablets with a reed pen.

121 **successively** : what comes after; forward into the future.

122 **Human Condition** : a standard default state of Human experience that is generally accepted to be the extent of its potential identity (*beingness*)—currently treated as *Homo Sapiens Sapiens,* but which is scheduled for replacement by *Homo Novus*.

123 **defragmentation** : the *reparation* of wholeness; a process of removing "*fragmentation*" in data or knowledge to provide a clear understanding; applying techniques and processes that promote a *holistic* interconnected *alpha* state, favoring observational *Awareness* of continuity in all spiritual and physical systems; in *NexGen Systemology*, a "*Seeker*" achieving an actualized state of basic "*Self-Honest Awareness*" is said to be *defragmented*.

In fact, these efforts have been going on for quite some time—so long, in fact, that it seems as if all the interested parties have already now arrived *here.*

In previous *Grade-III* instruction, a basic systemological interpretation of themes and events for the *Enuma Eliš* are provided to base descriptions and illustrate the Standard Model. This is of significant benefit to a Mardukite Seeker[124] continuing from the *Grade-II* "Mardukite Core."

At the completion of *Grade-III Mardukite Systemology*, a Seeker is expected to have mostly "flattened the waves"[125] that collapsed[126] around even those considerations fixed to the ancient Mesopotamian[127] semantics from which our Systemology is originally drawn. It may be very accurately stated that our applied Systemology is a further development of what was discovered in our revival of "Mardukite Zuism."[128] It directly prompted the discovery of an applied spiritual technology for the 21st Century AD that is clearly present during the 21st Century BC; from the time of the "Age of Aries" (c. 2160 B.C.) and the official Mardukite Babylonian systematization that has carried through to today.

Does this mean we are, in any way, rejecting the historical premise on which we first drew our knowledge?[129] ...certainly not. But, let us just say that it took thousands of years to uncover (or recover, depending on your perspective) and translate the cuneiform source of global cosmologies, mythologies and creation myths from ancient Babylon—the *Enuma Eliš*—and in more than a century since its widespread academic circulation in the late 1800's, it has still taken until *now* to develop any workable cohesion of its information for any effective spiritual ideal or application other than exploration of its cultural mythology as a series of

---

124 **Seeker** : an individual on the *Pathway to Self-Honesty*; a practitioner of *Mardukite Systemology* or *NexGen Systemology Processing* that is working toward *Ascension.*

125 **flattening a wave (processing-out)** : to reduce *emotional encoding* of an *imprint* to zero; to dissolve a *wave-form* or *thought-formed* "solid" such as a *"belief"*; to completely run a *process* to its end, thereby *flattening* any previously *"collapsed-waves"* or *fragmentation* that is obstructing the *clear channel* of *Self-Awareness*; also referred to as "processing-out"; to discharge all previously held emotionally encoded imprinting or erroneous programming and beliefs that otherwise fix the free flow (wave) to a particular pattern, solid or concrete *"is"* form.

126 **collapsing a wave (*wave-function collapse*)** : the definition or calculation of a wave-function or interaction of potential interactions by Observation. The idea that the Observer is collapsing the wave-function by measuring it appears in the field of *Quantum Physics*. Consciousness or *Awareness* "collapses" the wave-function of energy and matter as the third (required) Principle of Apparent Manifestation; potentiality as a wave is collapsed into an apparent *"is"* (which may be undone by "flattening" the wave back into its state of potentiality).

127 **Mesopotamia** : land between Tigris and Euphrates River; modern-day Iraq.

128 **Mardukite Zuism** : a Mesopotamian-themed (Babylonian-oriented) religious philosophy and tradition applying the spiritual technology based on *Arcane Tablets* in combination with "Tech" from *NexGen Systemology*; first developed in the New Age underground by Joshua Free in 2008 and realized publicly in 2009 with the formal establishment of the *"Mardukite Chamberlains."*

129 **knowledge** : clear personal processing of informed understanding; information (data) that is actualized as effectively workable understanding; a demonstrable understanding on which we may 'set' our *Awareness*—or literally a "know-ledge."

esoteric symbols and traditions. This is very much akin to the understanding and knowledge that hovers around that *first* level or "Gate" of realizations;[130] which we have markedly explored within parameters[131] of the *Grade-I Route of Magic & Mysticism*‡ with an intention that a Seeker will "flatten" the programming that keeps them suspended as the "effect" of that level of understanding.

During personal investigations into evolutions of Western mysticism, which led to the formal 2008 launch of Mardukite Ministries (Mardukite Zuism), one key avenue from *Grade-I* served as a greater platform then any other for an early precursor to our NexGen Systemology, notably referred to as "Druidism." This information is explored directly in our material for the *Grade-I Route of Druidism & The Dragon Legacy*.∞ It is actually on *this* very foundation, and explorations into the origins of ancient Druidism, that led the present author to decide upon Mesopotamia (and specifically Babylon) as a "public" emphasis for further esoteric developments—which were continued underground, by the Mardukite Chamberlains (Mardukite Research Organization), from 2009 until 2012, when the *Grade-II* "Mardukite Core" reached its completion; and the *second* "*Gate*" dislodged.

For the next eight years, "Mardukite Systemology"[132] developed quietly in the underground as a futurist or "NexGen" movement dedicated to achieving the "next step" on this *Pathway*—one that would inevitably up and out of the "systems" that had been laid out to entrap the occupation and ensnare the attentions, willpower[133] and spiritual energy of the true "Self" to this *beta-existence.* How then might we use the best of what we had found effective and

---

130 **realization** : the clear perception of an understanding; to make "real" or give "reality" to so as to grant a property of "beingness" or "being as it is"; the state or instance of coming to an *Awareness*; in *NexGen Systemology,* "gnosis" or true knowledge achieved during *systematic processing*; achievement of a new (or "higher") cognition, true knowledge or perception of Self in relation to reality.

131 **parameters** : a defined range of possible variables within a model, spectrum or continuum; the extent of communicable reach capable within a system or across a distance; the defined or imposed limitations placed on a system or the functions within a system; the extent to which a Life or "thing" can *be, do* or *know* along any channel within the confines of a specific system or spectrum of existence.

‡ The *Mardukite Grade-I Esoteric Research Library* for the "Route of Magick & Mysticism" consists of "*Arcanum: The Great Magical Arcanum*" by Joshua Free, in addition to "*Sorcerer's Handbook*" (first written as Merlyn Stone) and the anthology "*Vampyre's Handbook.*" (A Master Edition Hardcover anthology of these titles is expected in late 2020.)

∞ The *Mardukite Grade-I Esoteric Research Library* for the "Route of Druidism & The Dragon Legacy" consists of "*The Druid's Handbook,*" "*Elvenomicon*" and "*Draconomicon*" by Joshua Free, combined together (with additional material) for the 2020 Master Edition Hardcover anthology "*Merlyn's Complete Book of Druidism: A Master Course in Druidry for Modern Druids.*"

132 **NexGen Systemology** : a modern tradition of applied religious philosophy and spiritual technology based on *Arcane Tablets* in combination with "*general systemology*" and "*games theory*" developed in the New Age underground by Joshua Free in 2011 as an advanced futurist extension of the "*Mardukite Chamberlains.*"

133 **will** *or* **WILL** (5.0) : in *NexGen Systemology* (from the *Standard Model*)—the spiritual ability at (5.0) of an *Alpha Spirit* (7.0) to apply *intention* as "Cause" from a higher order of reasoning and consideration (6.0) than the thoughts found in *beta-existence,* where it manifests as "effect" below (4.0).

workable to return the individual Seeker toward the direction that they *truly* occupy as an "Alpha"[134] condition in a higher spiritual plane? This was no simple task; requiring *eight* dedicated years to underground research and experimentation.

A perceptive Seeker having followed the serpent trail through lower *Grades* will undoubtedly recognize many elements found in our Standard Model and "ZU-line" that we have perfected to represent a wide-encompassing[135] continuity of our Systemology (and Mardukite Zuism) as an effective workable paradigm. For example, the basic model structure is found consistent in both the ancient Babylonian sources *and* those qualifying a "Druid's Cabala" (from Welsh and other European sources).[*] The gradient distinction plotted as a "seven-plus-one" methodology[136] is mirrored in so many global representations along the timeline of human traditions—from the Eastern[137] *chakras*[138] to the *StarGates of Babylon*—that an entire volume could be prepared exclusively on the subject of such esoteric associations and correspondences;[139] information that an individual could otherwise spend their entire lifetime correlating[140] and still not reach any greater level of *realization* or higher point of *Actualized Awareness* that will carry them on upward toward an ideal state of *knowing* and *being*.

With the exception of a few semantics from Mesopotamia, we carry very little of the *"stuff"* along with us as we progress through higher gradients of understanding and personal *Awareness*. This has long been one of the shortcomings of previous attempts toward *Ascension*, whereby an initiate is not given tools to properly "let go" of the material programming and personal imprinting[141] along the way, and is instead forcibly applying untempered excessive effort to make a journey toward accumulation of "things" rather than a reduction; because

---

134 **alpha** : the first, primary, basic, superior or beginning of some form; in NexGen systemology, it also refers to the state of existence that operates on archetypes, will and intention "exterior" to the low-level condensation and solidarity of energy and matter as the 'physical universe'.

135 **encompassing** : to form a circle around, surround or envelop around.

* Many points of our Systemology may also be found among Welsh "Triad" teachings of Bards and Druids.

136 **methodology** : a system of methods, principles and rules to compose a systematic paradigm of philosophy or science.

137 **Eastern traditions** : the evolution of the *Ancient Mystery School* east of its origins, primarily the Asian continent, or what is archaically referred to as "oriental."

138 **chakra** : (an archaic term used by ancient wisdom traditions); an etheric wheel-mechanism that processes *ZU* energy at specific frequencies along the *ZU-line*, of which a Human being reportedly has *seven* at various degrees.

139 **correspondence** : a direct relationship or correlation; see also *"associative knowledge."*

140 **correlating** : the relationship between two or more aspects, parts or systems.

141 **imprint** : to strongly impress, stamp, mark (or outline) onto a softer 'impressible' substance; to mark with pressure onto a surface; in *NexGen Systemology*, the term is used to indicate permanent Reality impressions marked by frequencies, energies or interactions experienced during periods of emotional distress, pain, unconsciousness, loss, enforcement, or something antagonistic to physical (personal) survival, all of which are are stored with other reactive response-mechanisms at lower-levels of *Awareness* as opposed to the active memory database and proactive processing center of the Mind; an experiential "memory-set" that may later resurface—be triggered or stimulated artificially—as Reality, of which similar responses will be engaged automatically.

they have not experienced demonstrations of anything to the contrary. Yet, these former methods have not proved effective in producing anything more than a incredible collection of cliché axioms and spiritual doctrines that continue to keep the Human Condition in a fragmented state.

Within our Systemology, at each gradient of *Awareness,* a Seeker is systematically processed to "lighten their load" of *stuff,* because quite frankly, it will not all fit through as one moves further and further. And it was not meant to. Even the astral "levels" and energetic "layers" envisioned around the Alpha Spirit's consideration of a "finite body" should be *lessening,* not *increasing,* as one reaches towards *Infinity* on the *Pathway.* For this reason, many who have attempted "astral work" and "Gatewalking" (&tc.) in the past, and based on the esoteric instruction and other paradigms predating Mardukite Zuism and Systemology, have not found true successes toward the ultimate goal that could have otherwise been reached.

Many underground esoteric and mystic practitioners, that have known no better, *have* actually traversed the sevenfold system—but they have only done so from within the *first* sphere or "Gate," not realizing that the system repeats itself as a fractal[142] macrocosm[143] and microcosm in relative "directions" of magnification.* Most practitioners have either become lost to entanglement[144] of the "Gates" at a first level of understanding or end up abandoning the approach altogether.

Even many of the brightest and most aptly trained and skilled individuals in such practices have found themselves permanently encircling the first level of continuity with a genuine feeling that they have "arrived" and therefore tend to look no further, only fragmenting the continuity of what they have found into a greater amount of potential correspondences. This is because there *is* a continuity at each level of understanding whereby everything can be

---

142 **fractal** : a wave-curve, geometric figure, form or pattern, with each part representative of the same characteristics as the whole; any baseline, sequence or pattern where the 'whole' is found in the 'parts' and the 'parts' contain the 'whole'; a pattern that reoccurs similarly at various scales/levels on a continuous whole; a subset of a Euclidean space explored in higher-level academic mathematics, in which fractal dimensions are found to exceed topological ones; in NexGen Systemology, a "fractal-like" description is used specifically for a pattern or form that has a reoccurring nature without regard to what level or scale it is manifest upon. Examples include the formation of crystals, tree-like patterns, the comparison of atoms to solar systems to galaxies, &tc.

143 **macrocosmic** : taking examples and system demonstrations at one level and applying them as a larger demonstration of a relatively higher level or unseen dimension.

* This would equate to "7-times-7-plus-1" (49+1 or 50) individual "*Gates,*" "*systems,*" Divine "ME" or "*implants*" that a practitioner would have to appropriately "guess correctly" and adequately access and "flatten" if using the mystic-magician's "*Grade-I*" methodology. Naturally, very few have been successful in codifying such a pursuit.

144 **entanglement** : tangled together; intertwined and enmeshed systems; in *NexGen Systemology,* a reference to the interrelation of all particles as waves at a higher point of connectivity than is apparent, since wave-functions only "collapse" when someone is *Observing,* or doing the measuring, evaluating, &tc.

made to seem to fit within *that* potential level of knowledge[145] accessible from *that* POV[146] (viewpoint); just as much as we could restrict a total knowledge of purely physical phenomenon using a purely physical understanding of chemicals and forces and still be made to seem "correct" for *that* level of understanding and knowledge base.

More important than determining or demonstrating if any of our knowledge is representative of some "Absolute Truth," the emphasis of Systemology is toward that ancient lore which is found to be objectively[147] effective in predicting and workable in producing targeted results.

Our concern in presenting the "Standard Model"—and likewise why it is not introduced directly until *Grade-III*—pertains to previous associations[148] an individual tends to attempt to apply to this material as just more "esoteric lore" to incorporate into an existing databank. Our model is specific but representative; fluid as opposed to rigidly fixed; interconnected systematically rather than a compilation of parts treated in exclusion. All of the parts and facets and elements it represents, maintain a complete energetic circulation of communication with one another as a dynamic[149] system; a system that is always changing, shifting and altering its apparent[150] face; and hence why these systems continue to persist with any solidity.

An individual's apparent personal stability is often based on the conditional "agreements" they have made concerning Reality—which is to say a determination[151] of considerations

---

145 **knowledge** : clear personal processing of informed understanding; information (data) that is actualized as effectively workable understanding; a demonstrable understanding on which we may 'set' our *Awareness*—or literally a "know-ledge."

146 **point-of-view (POV)** : an opinion or attitude as expressed from a specific identity-phase; a specific standpoint or vantage-point; a definitive manner of consideration specific to an individual phase or identity; a place or position affording a specific view or vantage; circumstances and programming of an individual that is conducive to a particular response, consideration or belief-set (paradigm); a position (consideration) or place (location) that provides a specific view or perspective (subjective) on experience (of the objective).

147 **objectively** : concerning the "external world" and attempts to observe Reality independent of personal "subjective" factors.

148 **associative knowledge** : significance or meaning of a facet or aspect assigned to (or considered to have) a direct relationship with another facet; to connect or relate ideas or facets of existence with one another; a reactive-response image, emotion or conception that is suggested by (or directly accompanies) something other than itself; in traditional systems logic, an equivalency of significance or meaning between facets or sets that are grouped together, such as in *(a + b) + c = a + (b + c)*; in NexGen Systemology, erroneous associative knowledge is assignment of the same value to all facets or parts considered as related (even when they are not actually so), such as in *a = a, b = a, c = a* and so forth without distinction.

149 **dynamic (systems)** : a principle or fixed system which demonstrates its *'variations'* in activity (or output) only in constant relation to variables or fluctuation of interrelated systems; a standard principle, function, process or system that exhibits *'variations'* and change simultaneously with all connected systems.

150 **apparent** : visibly exposed to sight; evident rather than actual, as presumed by Observation; readily perceived, especially by the senses.

151 **Self-determinism** : the freedom to act, clear of external control or influence; the personal control of Will to direct intention.

about what is *real*. At its core, this is actually so important, that any useful meaning it might have carried was lost to the cliché sentiment that "everyone creates their own reality." But such statements have done nothing to effectively and successfully liberate considerations of the *Self* from its material entrapment. We tend to speak of "well-adjusted" individuals quite simply as those that seem to face new data and experiences *anew*, without overly fixating, comparing or associating past data and experience.[152] This, in itself, is milestones ahead from the standard issue[153] state of the Human Condition.

Therefore, we have found, as a basic barrier to increasing an individual's *Awareness*—and as a basis of the "problems" facing the Human Condition—the inability to adjust significances and reassign "importances" for new evaluations,[154] while simultaneously under the hold and command of reactive-response programming and other heavily imprinted (or energetically charged)[155] past experiences. As this personal inability continues to be validated,[156] presumably across multiple lifetimes, the Self finds itself becoming more and more the "effect" of fixed mental implants and finite considerations of reality—and thus we find ourselves now stuck here, as the ultimate result of trillions of Alpha Spirits all succumbing[157] to the same downward spiral of considerations and manifestation,[158] imprisoned in a very solid Physical Universe.

---

152 **experiential data** : accumulated reference points we store as memory concerning our "experience" with Reality.

153 **standard issue** : equally dispensed to all without consideration.

154 **evaluate** : to determine, assign or fix a set value, amount or meaning.

155 **charge** : to fill or furnish with a quality; to supply with energy; to lay a command upon; in *NexGen Systemology*—to imbue with intention; to overspread with emotion; application of *Self-directed (WILL)* "intention" toward an emotional manifestation in beta-existence; personal energy stores and significances entwined as fragmentation in mental images, reactive-response encoding and intellectual (and/or) programmed beliefs; in traditional mysticism, to intentionally fix an energetic resonance to meet some degree, or to bring a specific concentration of energy that is transferred to a focal point, such as an object or space.

156 **validation** : the reinforcement of agreements of Reality.

157 **succumb** : to give way, or give in to, a relatively stronger superior force.

158 **manifestation** : something brought into existence.

:: 2 ::

UNIVERSAL COMMUNICATION, CONTROL AND COMMAND
« SYSTEMOLOGY GRADE-IV CRASH COURSE[159] »

Systemology of "Communication, Control & Command" is a primary emphasis of our *Grade-IV* material for Metahuman Systemology—both for individual "Seekers" *and* those practicing our systematic processing[160] methodology as "Professional Pilots"[161] (and "ministers" of Mardukite Zuism). A complete course on these subjects is presented within two volumes: *"Communication and Control of Energy & Power"* (*Liber-2C*) and *"Command of the Mind-Body Connection"* (*Liber-2D*). For our present purposes—to both newcomers and returning Mardukite Systemologists alike—a crash course on the fundamentals will suffice in carrying the spirit of this present *"Liber-3C"* manual.

The Standard Model of Systemology demonstrates a vast network of communication between our proposed points of "zero" and "Infinity"—most of which, as it relates to the individual themselves, is experienced along a personal energetic continuum of potential "beingness" that we call the ZU-line. Combined, the two "concepts" represent all possible interactive points between an individual and a universe—*any* universe.

> The relay of energy, a message or signal—or even locating a personal POV (viewpoint) for the Self—along this continuum is referred to as <u>communication</u>.

> Communication relayed from an operative center or organizational cluster, which incites[162] new activity elsewhere on the ZU-line, is considered <u>control</u>.

> Abilities of the Self (I-AM), from its ideal exterior POV as Alpha Spirit, to direct a communication for control that is perfectly duplicated along the ZU-line without fragmentation is true <u>command</u>.

From a systematic approach, *communication* is the primary catalyst[163] by which all *Life* is learning and experiencing existence. We are directing and receiving communications from the external environment while interacting with the Physical Universe (*beta-existence*). These are all processed by communications internal and interior to the Mind–Body connection,

---

159 **crash-coursed** : a very intense or steep delivery of education over a very brief time period.

160 **processing, systematic** : the inner-workings or "through-put" result of systems; in *NexGen Systemology*, a methodology of applied spiritual technology used toward personal Self-Actualization; methods of selective directed attention, communicated language and associative imagery that targets an increase in personal control of the human condition.

161 **pilot** : the steersman of a ship; in *NexGen Systemology*—an individual qualified to operate *Systemology Processing* for other *Seekers* on the *Pathway to Self-Honesty*.

162 **incite** : to urge on or cause; instigate; prove or stimulate into action.

163 **catalyst** : something that causes action between two systems or aspects, but which itself is unaffected as a variable of this energy communication; a medium or intermediary.

upon which experience and command of the Human Condition seems primarily anchored.[164]

Humanity has run through many phases of intellectual reach to properly resume this control, ever since perfected knowledge of the Mind–Body connection became fragmented thousands of years ago. Yet, there is an inherent *knowing* that behind, back of, and beneath the "surface" of what we are consciously facing as reality in this Physical Universe, there is an entire existence that is blocked, occluded, occulted[165] or otherwise obscurely hidden from Human view, if following along to the beat of standard issue programming.

Physical sciences have offered little more than further "agreements"[166] to confine our considerations of thought and spirit to this Physical Universe. Eventually an individual finds that all conceptions of potential beingness are either "reactive" or else tied strongly to "mental programming implants" that selectively direct and fix our attentions on the lowest denominator of beta-existence; that which we refer to as the "RCC"[167] or "Reactive Control Center" in Systemology, which generates bio-chemical and emotional experiences internally in a physical body ("genetic vehicle").

The purpose of any esoteric initiatory or mystical gradient system that mirrors facets of the "Gates" or "Levels" (which we demonstrate as a "zero-to-eight scale" on the Standard Model) were originally intended to systematically and progressively remove the standard issue programming, emotional encoding[168] and other implants that had been taken on, reinforced and validated during the course of an exceptionally long spiritual existence.

---

164 **anchor (*conceptual*)** : a stable point in space; a fixed point used to hold or stabilize a spatial existence of other points; a spatial point that fixes the parameters of dimensional orientation, such as the corner-points of a solid object in relation to other points in space; in *NexGen Systemology*, "beta-anchored" is an expression used to describe the fixed orientation of a viewpoint from Self in relation to all possible spatial points in *beta-existence* ("physical universe"), or else the existential points that fix the operation of the "body" within the space-time of *beta-existence*.

165 **eclipse / occult** : to cast a shadow or darken; to block out or obscure a comparison.

166 **agreement** : unanimity of opinion; an accepted arrangement; "reality."

167 **reactive control center (RCC)** : the secondary (reactive) communication system of the "*Mind*"; a relay point of *Awareness* along the Identity's *ZU-line*, which is responsible for engaging basic motors, biochemical processes and any *programmed automated responses* of a living *beta* organism; the reactive Mind-Center of a living organism relaying communications of *Awareness* between causal experience of *Physical Systems* and the "*Master Control Center*"; it presumably stores all emotional encoded imprints as fragmentation of "chakra" frequencies of *ZU* (within the range of the "*psychological/emotive systems*" of a being), which it may *react* to as Reality at any time; in *NexGen Systemology*, this is plotted at (2.0) on the continuity model of the *ZU-line*.

168 **emotional encoding** : the substance of *imprints*; associations of sensory experience with an *imprint*; perceptions of our environment that receive an *emotional charge*, which form or reinforce facets of an *imprint*; perceptions recorded and stored as an *imprint* within the "emotional range" of energetic manifestation; the formation of an energetic store or charge on a channel that fixes emotional responses as a mechanistic automation, which is carried on in an individual's spiritual timeline or personal continuum of existence.

By reducing the weight of these lower level energy masses from the "banks" of the "spirit," an initiate was treated to a progressive journey toward a greater metahuman POV that put them back in contact with the ZU-line from their true and ideal state.

While a methodology of using "Gates" proved successful in the beginning (many thousands of years ago), these organized efforts to spiritually liberate individuals from the material system trappings of this Physical Universe (Earth-Gate or Zero-Gate) did not continue undefiled[169] for very long. In no short time thereafter, we find the clear path obscured and confounded into "*Mystery Traditions*" with a now fragmented knowledge dispersed across varying cultures throughout the globe. This is the true nature of the "Tower of Babylon Incident" whereby complete, clear and present access to the "Gates of Understanding" was cut off from humanity.[*]

In Systemology we do more than just suppose there is more than inert material continuity of this Physical Universe—we go forth to codify[170] and systematize the understanding available to us on the Standard Model. As such, we have noted the existence of the "RCC" and "MCC"[171] in previous texts, plotted at "2.0" and "4.0" respectively on the ZU-line. This knowledge is a primary emphasis of education and systematic processing demonstrations provided in *Grade-III*. What we have done on the Standard Model is provided a structure for the Mind–Body connection that, when operated by a Self-Honest[172] individual, is under the command of Self as the Alpha Spirit, free of entrapment to low-level considerations and automated reactivity to the environment.

Our initial presentation of the Standard Model in *Grade-III*, introduced this systematic structure by focusing on the "positive" end of the scale—that which can be brought to complete visibility and scrutiny by the *Actualized Awareness* of the Alpha Spirit (*Self*).

---

169 **undefiled** : to remain intact, untouched or unchanged; to be left in an original "virgin" state.

* See also "*Tablets of Destiny*" (*Liber-One*) and various information dispersed throughout the Grade-II "Mardukite Core."

170 **codification** : the process of arranging knowledge in a systematic form.

171 **master control center (MCC)** : a perfect computing device to the extent of the information received from "lower levels" of sensory experience/perception; the proactive communication system of the "*Mind*"; a relay point of active *Awareness* along the Identity's *ZU-line*, which is responsible for maintaining basic *Self-Honest Clarity* of *Knowingness* as a *seat of consciousness* between the *Alpha-Spirit* and the secondary "*Reactive Control Center*" of a *Lifeform* in *beta existence*; the Mind-center for an *Alpha-Spirit* to actualize cause in the *beta existence*; the analytical *Self-Determined* Mind-center of an *Alpha-Spirit used* to project *Will* toward the genetic body; the point of contact between *Spiritual Systems* and the *beta existence*; presumably the "*Third Eye*" of a being connected directly to the *I-AM-Self*, which is responsible for *determining* Reality at any time; in *NexGen Systemology*, this is plotted at (4.0) on the continuity model of the *ZU-line*.

172 **Self-honesty** : the *alpha* state of *being* and *knowing*; clear and present total *Awareness* of-and-as *Self*, in its most basic and true proactive expression of itself as *Spirit* or *I-AM*—free of artificial attachments, perceptive filters and other emotionally-reactive or mentally-conditioned programming imposed on the human condition by the systematized physical world.

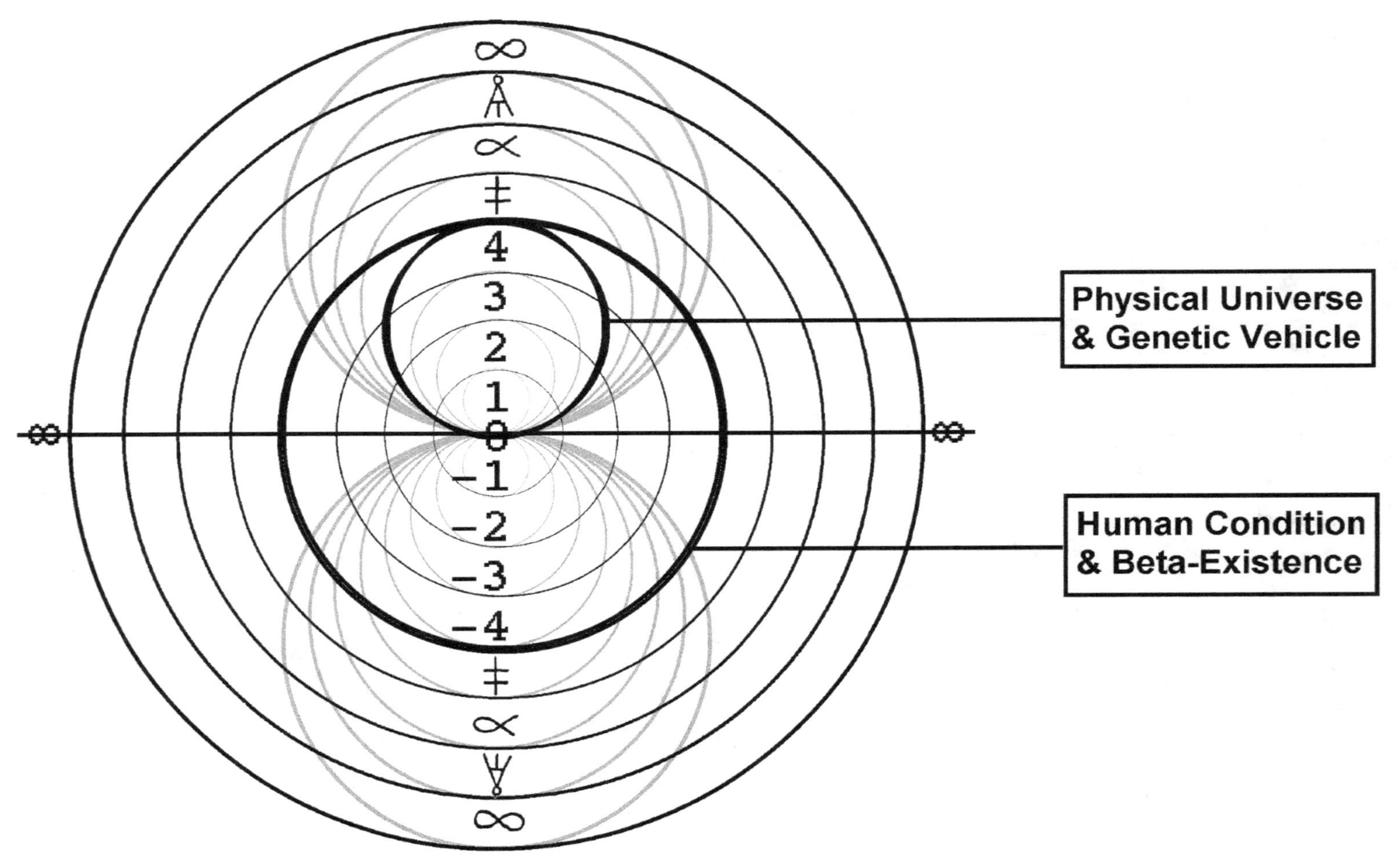

Physical Universe & Genetic Vehicle
Human Condition & Beta-Existence

What we are here describing, between "0.1" and "4.0" is the "internal" nature of the Mind–Body connection, as fully within the grasp of personal experience, as one interacts with the "spheres of existence," but particularly in direct contrast to the "external" qualities of the Physical Universe.

It is at the zero-point of our model that the most "solid" aspects of the "genetic vehicle" meet or match frequencies of "solid matter" in this Physical Universe. We tend to treat the entire range of "0.0" to "4.0" as *beta-existence*, because it reflects the total scope of physical, emotional and mental parameters as experienced "internally" from the POV of an Alpha Spirit operating its *beingness* "within" physical conditions of a "genetic vehicle"—and this is hardly an optimum[173] position of command for a God-like spiritual being.

> (0.0) : "External" (continuity of objective beta-existence; Physical Universe)
>
> (0.1) to (4.0) : "Internal" (*Self*, operating as POV in a "genetic vehicle")

Mardukite Systemology *Grade-III*, as upper-most of our "Master" Grades,[174] refined the levels of knowledge within its domain toward operation of the Human Condition based on the "internal" systems most readily accessible by a Seeker—even without professional "Piloted" assistance. As we began to refine the "Wizard" levels[‡] toward higher spiritual pursuits, it became clear that we would be facing the other balancing polar end of this spectrum;[175] that which contained the very "interior" root of all encoding, imprinting, mental implants and imagery that we were otherwise interacting with "internally."

To accomplish goals of accessing these higher *realizations* to continue our graded *Pathway*, it became necessary to codify and systematize a wider-angle view of our model that would resolve the "problems" that we were left with in processing toward a *Homo Novus*[176] "metahuman" state.

This new Wizard-level work continues to beoperated and developed—then "processed"—on

---

173 **optimum** : the most favorable or ideal conditions for the best result; the greatest degree of result under specific conditions.

174 **"Master Grades"** : literary materials by Joshua Free (written between 1995 and 2019) revised and compiled for the International School of Systemology instructional grades—"Route of Magick & Mysticism" (*Grade I, Part A*), "Route of Druidism & Dragon Legacy" (*Grade I, Part D*), "Route of Mesopotamian Mysteries" (Grade II) and "Route of Mardukite Systemology" (*Grade III*), collectively known as the "Pathway to Self-Honesty."

‡ *Grades IV–VII = Wizard Levels 0–III*

175 **spectrum** : a broad range or array as a continuous series or sequence; defined parts along a singular continuum.

176 **Homo Novus** : literally, the "new man"; the "newly elevated man" or "known man" in ancient Rome; the man who "knows (only) through himself"; in NexGen Systemology—the next spiritual and intellectual evolution of *homo sapiens* (the "modern Human Condition"), which is signified by a demonstration of higher faculties of *Self-Actualization* and clear *Awareness*.

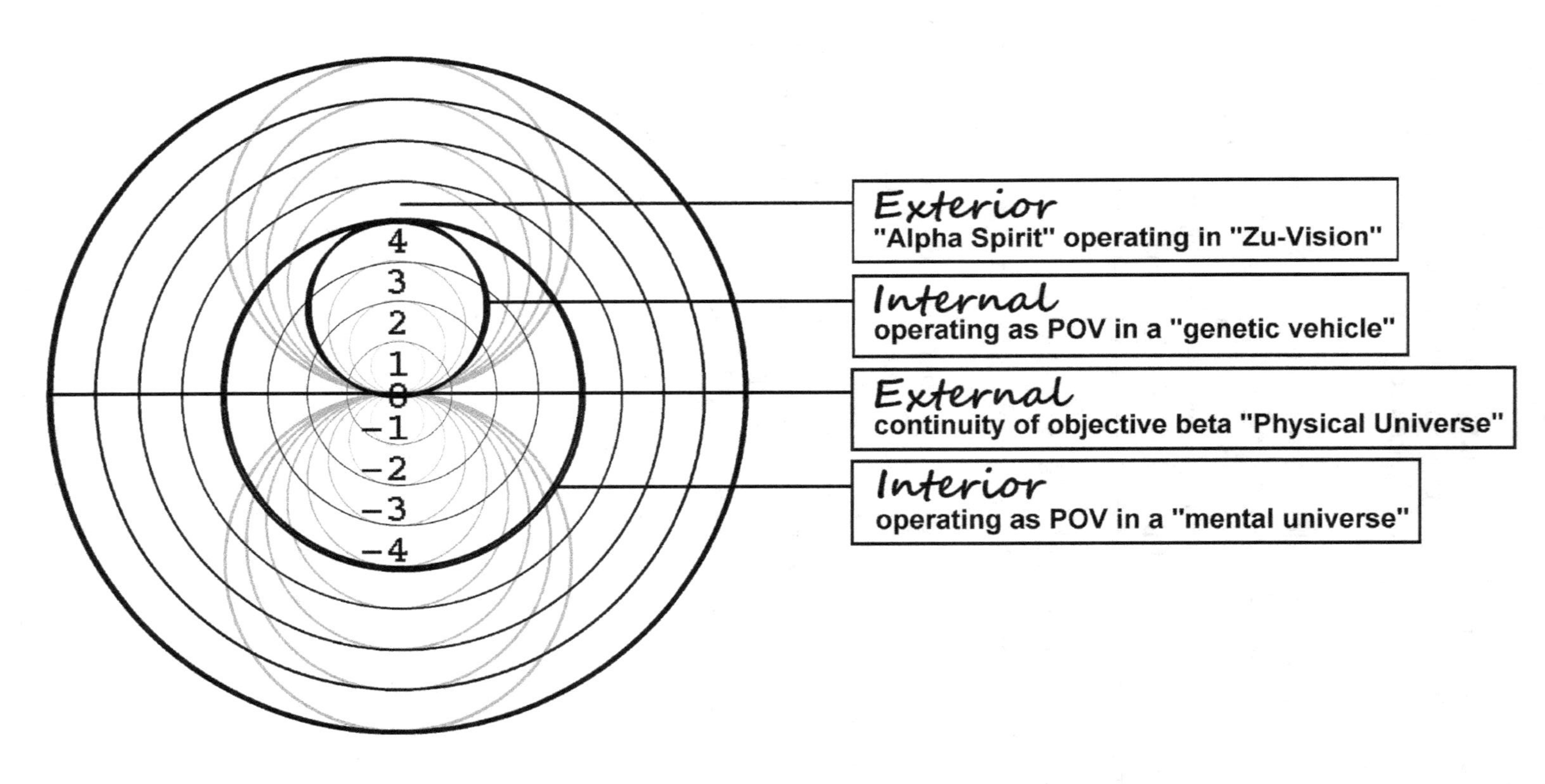

4
3
2
1
0
-1
-2
-3
-4
Exterior
"Alpha Spirit" operating in "Zu-Vision"
Internal
operating as POV in a "genetic vehicle"
External
continuity of objective beta "Physical Universe"
Interior
operating as POV in a "mental universe"

a smooth gradient, just as we find delivered in "*Crystal Clear*" (*Liber-2B*) and the advanced professional piloting procedures provided in the appendix of "*Command of the Mind–Body Connection*" (*Liber-2D*) as it relates to Master Grade "processing."

What we are simply concerned with now is a further reach that extends our progression on this *Pathway*, beyond simply the "internal" workings of the genetic vehicle we are operating, but the communications throughout the "interior" of our personal "mental universe"— which includes the Mind–Body connection (between "0.1" and "4.0") *in addition to* the full "interior" of the personal Mind-System (even independent of a specific "physical body") carried on an individual's persona energetic continuum (or ZU-line) all the way down to the (sub?) range of "–4.0" on our Standard Model.

This means that using our systematic reasoning, the full operating system of the "Mind" exists for each individual as a "mental universe system" that extends from "4.0" to "–4.0" on our Standard Model demonstration. That entire zone of existence is marked as "interior" (using our semantics), because the *beingness* of an individual is still operating from a POV "interior" to the mental systems—particularly "beta" mental systems—thus it's stated that the individual is *still* very much "in their head" (figuratively speaking). This, of course, adds another dynamic to our model.

(4.0) to (0.1) : "Internal" (*Self*, operating as POV in a "genetic vehicle")

(4.0) to (–4.) : "Interior" (*Self*, operating as POV in a "mental universe")

It became evident, as we plotted out the work for the Wizard Grades, that we would have to make certain of our distinctions regarding these classifications on the Standard Model; particularly the differences between the POV states of *Self* that are "internal," "interior" or "exterior"—because it becomes quite relevant when accessing the upper-routes.

This whole matter was disregarded as fancy word play and semantic tricks for many years until we realized that all previous "traditional" attempts at metahuman states relying on lower-Grade instructions regarding "astral vision" and "spirit bodies" *did not* actually provide effective tools on getting an individual *exterior* to the "mental universe." And without systematic processing to defragment mental programming, *no one ever has* (using these former esoteric, mystic, spiritual and religious methods).

These challenges were met and resolved, though it is important for the Seeker to at least understand how these methods, techniques and the Standard Model that they are drawn from, all came together in a very logical and systematic way that relied on very few assumptions; rather, it has been drawn up out of necessity, demonstrating the "*simplest*" that these systems could be to exist. There are no doubt a near-infinite number of variations, fragments and smaller operating systems that could be plotted using this information. Many of the larger implants and prominent micro-systems are directly encountered and processed as we move upward in these higher Grades through the *Gateways to Infinity*.

∆ ∆ ∆ ∆ ∆ ∆ ∆

Based on the logics presented in *Grade-III Mardukite Systemology*, it has been established that the root common denominator behind all of existence, which is to say its "prime directive," is *to exist*. It cannot be stated any simpler as the problem and its answer or solution are one and the same. As it turns out, nearly all the pre-programmed (implanted) networks of the Mind that are linked to what we might call "human problems" are of this exact same nature.

Given the tremendous power of the Mind, it would seem logical that an individual would use these functions and creative faculties[177] to achieve the ideal state of *beingness*, the optimum level of *existence*. And it has been found that, at the core, this is the creative intent and purpose behind the goals of the Alpha Spirit way back in the beginning of the personal timeline. In the original and true spiritual state of "I-AM" (7.0), the Alpha Spirit is operating at its highest efficiency within close proximity to "Source" (8.0).

The abilities of the Alpha Spirit throughout its own creative journey are linked precisely to the communication systems that are demonstrated on the *Zu-line* of the Standard Model. We see that when an individual is operating from their "MCC"—represented with the symbol of the upward pointing triangle—they are "facing up" or "confronting"[178] the reality of their universe, reaching and extending across spheres of influence in the existential[179] or objective universe, and is learning from the association and incorporation of its own experiential knowledge. In Systemology, we say that such an individual is operating "analytically" from some degree between "2.1" and "4.0" on our model; which is in contrast to the "reactive mechanisms" and "emotional responses" governed by the "RCC" at lower degrees of beta-Awareness[180]—and all of this is quite adequately demonstrated in our *Grade-III* literature and equally contacted with the supplemental "processing" suggestions prior to this present publication.

In our previous manuals of instruction on systematic processing, a *Seeker* or *Pilot* is dealing primarily with *"products of"* the Mind-System when treating resulting conditions found within the "internal" systems. It was not until we began to systematize our knowledge of

---

177 **faculties** : abilities of the mind (individual) inherent or developed.

178 **confront** : to come around in front of; to be in the presence of; to stand in front of, or in the face of; to meet "face-to-face" or "face-up-to."

179 **existential** : pertaining to existence, or some aspect or condition of existence.

180 **beta (awareness)** : all consciousness activity ("*Awareness*") in the "Physical Universe" (KI) or else *beta-existence*; *Awareness* within the range of the *genetic-body*, including material thoughts, emotional responses and physical motors; personal *Awareness* of physical energy and physical matter moving through physical space and experienced as "time"; the *Awareness* held by *Self* that is restricted to a physical organic *Lifeform* or "*genetic vehicle*" in which it experiences causality in the *Physical Universe*.

how the emotional encoding,[181] imprinting and other programming, actually takes place and the way in which it is even stored between lifetimes as part of the spiritual identity,[182] that we realized that there was a more deeply ingrained chain of potential "terminals"[183] or "nodes" by which all of these later programs, tendencies, fixations, compulsions,[184] avoidances (*&tc.*) could even attach to an individual in any way. Since a few of us began to call these types of interior facets "implants" early on in our research, the name stuck.

On an energetic level—whether physical kinetics,[185] emotional charges and mental circuits[186] —we find varying "lines" and "connections" formed with various "terminals" of existence that we may have a communication with. Many Seekers discover that they have a great many "ties" with various objects and people; but specifically as a representative symbol that is interacted with in *beta existence.*[187] These same "terminals" may be contacted or envisioned internally using mental faculties just as they are sprung up on automatic as "screens" to our view, whenever they are triggered or stimulated by the environment.

---

181 **emotional encoding** : the substance of *imprints*; associations of sensory experience with an *imprint*; perceptions of our environment that receive an *emotional charge*, which form or reinforce facets of an *imprint*; perceptions recorded and stored as an *imprint* within the "emotional range" of energetic manifestation; the formation of an energetic store or charge on a channel that fixes emotional responses as a mechanistic automation, which is carried on in an individual's spiritual timeline or personal continuum of existence.

182 **identity** : the collection of energy and matter—including memory—across the *"Spiritual Continuum"* that we consider as "I" of *Self.*

183 **terminal (node)** : a point, end or mass on a line; a point or connection for closing an electric circuit, such as a post on a battery terminating at each end of its own systematic function; any end point or 'termination' on a line; a point of connectivity with other points; in systems, any point which may be treated as a contact point of interaction; anything that may be distinguished as an 'is' and is therefore a 'termination point' of a system or along a flow-line which may interact with other related systems it shares a line with; a point of interaction with other points.

184 **compulsion** : a failure to be responsible for the dynamics of control—starting, stopping or altering—on a particular channel of communication and/or regarding a particular terminal in existence; an energetic flow with the appearance of being 'stuck' on the action it is already doing or by the control of some automatic mechanism.

185 **kinetic** : pertaining to the energy of physical motion and movement.

186 **circuit (feedback loop)** : a complete and continuous circuit flow of energy or information directed as an output from a source to a target which is altered and return back to the source as an input; in *General Systemology*—the continuous process where outputs of a system are routed back as inputs to complete a circuit or loop, which may be closed or connected to other systems/circuits; in *NexGen Systemology*—the continuous process where directed *Life* energy and *Awareness* is sent back to *Self* as experience, understanding and memory to complete an energetic circuit as a loop.

187 **beta (existence)** : all manifestation in the "Physical Universe" (KI); the "Physical" state of existence consisting of vibrations of physical energy and physical matter moving through physical space and experienced as "time"; the conditions of *Awareness* for the *Alpha-spirit* (*Self*) as a physical organic *Lifeform* or "*genetic vehicle*" in which it experiences causality in the *Physical Universe.*

It has been realized that the average individual carries a great deal of energetic "charge"[188] on their personal "mental images" which are treated as a reality substitution for the objective universe. In brief, the individual is interacting and reacting based on the mental and emotional stores connected to a "terminal" rather than the objective nature of the "thing." An individual goes as far as to "create" their copy of the objective universe based on automatic mechanisms and even begins to take for granted the concept that walls and other solids are *more real* than anything that could be created by the Self. When the individual has ceased to consciously create, they have succumbed to considerations that they are simply an effect. Naturally, the more numerous the strong "ties" to terminals in the Physical Universe, the stronger the "pull" to remain at such a level in order to receive whatever the individual is now wired to experience as an effect.

Using "*Systematic Operating Procedure 2-C*,"‡ a professional Systemology Pilot is trained and skilled to assist a Seeker in systematically "flattening the waves"[189] of collapsed energetic charges and masses that the individual is carrying within them on their personal identity continuum (*ZU-line*).[190] There are other ways explored later in *Grade-IV* that include the handling of "mental imagery" directly, but at this initial arrival to the fourth *Gate* or *Grade* of "understanding," we are primarily focused on the clear communication relays between all concepts and terminals that are easily accessible using these methods. Such techniques tend to be "generalized" in their approach so that a Seeker may insert their own applicable examples and yet still arrive at the ultimate conclusion or *realization* that each of the systematic processes are intended to achieve.

"*Analytical Recall*" (AR-SP) or "*Route-2*" techniques are introduced in "*Crystal Clear*" (*Liber-2B*) because they are the most appropriate for "Self-Processing"—which was the primary intent of *Grade-III*. However, in *Grade-IV* we have introduced a new organizational flow to our methods that involves "Piloted Processing" by a professional. This does not eliminate the possibilities of a Seeker working through the upper-routes of the Wizard levels on their own;

---

188 **charge** : to fill or furnish with a quality; to supply with energy; to lay a command upon; in *NexGen Systemology*—to imbue with intention; to overspread with emotion; application of *Self-directed (WILL)* "intention" toward an emotional manifestation in beta-existence; personal energy stores and significances entwined as fragmentation in mental images, reactive-response encoding and intellectual (and/or) programmed beliefs; in traditional mysticism, to intentionally fix an energetic resonance to meet some degree, or to bring a specific concentration of energy that is transferred to a focal point, such as an object or space.

‡ Introduced in "*Communication and Control of Energy & Power*" (*Liber-2C*) and refined in "*Command of the Mind-Body Connection*" (*Liber-2D*).

189 **flattening a wave (processing-out)** : to reduce *emotional encoding* of an *imprint* to zero; to dissolve a *wave-form* or *thought-formed* "solid" such as a "*belief*"; to completely run a *process* to its end, thereby *flattening* any previously "collapsed-waves" or *fragmentation* that is obstructing the *clear channel* of *Self-Awareness*; also referred to as "processing-out"; to discharge all previously held emotionally encoded imprinting or erroneous programming and beliefs that otherwise fix the free flow (wave) to a particular pattern, solid or concrete "*is*" form.

190 **identity-system** : the application of the *ZU-line* as "I"—the continuous expression of *Self* as *Awareness*.

but due to the caliber of work involved at these higher grades of development, the standard traditional delivery has always been confined to elite orders, societies, schools and "wizard academies" operating under the directed guide of those that have "already been where you want to go."

"Processing" or "systematic processing" that we "run" in our Systemology is composed of PCLs[191]—"Processing Command Lines"—that function very similarly on the Mind-System as you might operate a computer. Of course, this actually requires selective directed attention of the Mind, and the Seeker's willingness[192] to provide *presence*[193] to the "processing session." Methods for encouraging this, in addition to a complete course on the "Systemology of Communication, Control & Command" is provided in previous *Grade-IV* manuals.[‡]

Using an application of *Grade-III* tech, as even appropriate for basic Self-Processing, a Seeker would apply the PCL, for example: "What are *you willing* to *communicate* to another?" And this would be applied repeatedly, noting of course, that an individual will naturally "process" what is nearest or more readily available to them first; which also includes those concepts and terminals that validate personal safety and the "certainties" that they *know*. This is an excellent first step toward clearing lines of communication on the channels[194] that we find circuiting all throughout the "interior" of the Mind-System. When delivered correctly, the PCL eventually leaves the Seeker to *realize* that *they* are determining the extent to which they would be *willing* to reach (communicate). Once that is resolved, the Seeker or Pilot simply moves down to the next PCL or series on the list.[∞]

Two variations on this basic processing methodology are introduced in *Grade-IV*: "terminals" and "circuits."[195] If we use the above example, then some various "terminals" that could apply, include "phases"[196] and "roles" of individuals we have encountered in our lives, such

---

191 **processing command line (PCL)** or **command line** : a directed input; a specific command using highly selective language for *Systemology Processing*; a predetermined directive statement (cause) intended to focus concentrated attention (effect).

192 **willingness** : the ability and consideration to reach, face or confront some thing or energy; the ability and consideration to communicate along some line to produce an effect, to put attention or intention on the line.

193 **presence** : personal orientation of Self located in space and time and handling the energy-matter present; the quality of some thing (energy/matter) being "present" in space-time.

‡ *"Communication and Control of Energy & Power"* (*Liber-2C*) and refined in *"Command of the Mind-Body Connection"* (*Liber-2D*).

194 **channel** : a specific stream, course, direction or route.

∞ See the appendix of *"Command of the Mind-Body Connection"* (*Liber-2D*).

195 **circuit (feedback loop)** : a complete and continuous circuit flow of energy or information directed as an output from a source to a target which is altered and return back to the source as an input; in *General Systemology*—the continuous process where outputs of a system are routed back as inputs to complete a circuit or loop, which may be closed or connected to other systems/circuits; in *NexGen Systemology*—the continuous process where directed *Life* energy and *Awareness* is sent back to *Self* as experience, understanding and memory to complete an energetic circuit as a loop.

196 **phase** : in *NexGen Systemology,* a pattern of personality or identity that is assumed as the POV from *Self;*

as "mother" and "father" or even in greater society, such as "teacher" or "priest" *&tc.* It is presumed that a Seeker is going to find that some such possible terminals actually carry heavy energetic charges—which is to say that contact with (including a mental image) incites automatic reactionary responses. The encoding is negated when intentionally[197] processing with personal determinism rather than allowing them to go on being validated as a reaction that ensues outside the command of *Self*.

Further research and experimentation with these terminal-nodes, and connectivity of communication maintained along the ZU-line, resulted in their treatment as "circuits" in order to ensure that these PCLs were utilized to the furthest extent of their effectiveness. This meant approaching at least three primary energetic flows[198] of communication that contribute to the encoding, imprinting and programming that embeds itself on deeply seeded implants that are, in essence, "hardwired" to receive and store specific types of information. Therefore "Route-3" is so named, not only because it is our third standard procedure, but because it emphasizes recall and/or consideration on three distinct communication circuits between *Self* and any terminal.[*]

> Circuit-1/*outflow* : Self to another/others ("I" → X)
>
> Circuit-2/*inflow* : another/others to Self ("I" ← X)
>
> Circuit-3/*crossflow* : another/others to another/others ("I"; X → Y)

What this essentially did for our systematic processing was expedite the results or achievement of realizations by resolving what an individual "carries" with them by treating it from all angles. In the example previously given regarding the willingness to communicate; a standard PCL such as that be treated as "Circuit-1" (by *Grade-IV* standards) and applicably followed up with "What communication would you be willing to receive from X?" (Circuit-2) and "What communications would X be willing to share with Y?" Eventually the Seeker gets a good handle on their Self-determined management of "willingness to communicate." And this is only one facet regarding the subject of communication; and there are many. And communication itself being only one facet of the total *Pathway*, but a critical and fundamental one for achieving any further success later on.

---

see also *"phase alignment."*

197 **intention** : the directed application of Will; to intend (have "in Mind") or signify (give "significance" to) for or toward a particular purpose; in *NexGen Systemology* (from the *Standard Model*)—the spiritual activity at WILL (5.0) directed by an *Alpha Spirit* (7.0); the application of WILL as "Cause" from a higher order of Alpha Thought and consideration (6.0), which then may continue to relay communications as an "effect" in the universe.

198 **flow** : movement across (or through) a channel (or conduit); a direction of active energetic motion typically distinguished as either an *in-flow, out-flow* or *cross-flow.*

* An advanced variation of "Route-3" for A.T. (*Actualized Technician*) or "Wizard" use includes a fourth circuit, referred to as "Circuit-0" emphasizing the interior energetic communication flow of *"Self-to-Self."*

△ △ △ △ △ △ △

One of the benefits to Master Instruction of the lower *Grades* on the *Pathway to Self-Honesty* and Professional Piloted assistance for the upper-routes through the *Gateways to Infinity* is that expert attention can be given to an individual and their case. Very often these manuals are linked to information provided in other manuals; and as a whole, the entire knowledge-base has been built upon quite an extensive "esoteric research library" provided as the Master Grades.

It is important that a Seeker is not left to wander about too long on their own or become discouraged by their attempts to take very impersonal deliveries of "book-learning" and not necessarily arrive at the indicated destinations on their own.

This essential premise that we began with is very simple to consider. Systemologists learn to think systematically and apply the fractal-like gradient scale of the Standard Model to daily life—and those training to be "pilots" and "ministers" within our tradition go on to expertly apply these same elements to our processing procedures, which are systematically designed to return command of the Human Condition to *Self*, rather than operating from a POV that is controlled by these lower energy-driven mechanisms.

Connecting the practices to various "Gates" and levels of actualization[199] was simply a matter of expert observation by those of us who have forged ahead to clear the way; ensuring the certainty that each level of instruction and practice is undoubtedly *leading to* a further progression on the *Pathway.*

In the example provided concerning basic communication processes: even the subject or term "communication" as a general concept may be too far of a reach for someone just starting out. But of course the "books" will only provide the most widely-useful or effective PCL (based on our own personal research and discoveries in the offices). In this instance, a *Pilot* recognizing this, would engage in communication about the terminal; it may be found that the most appropriate way of approaching this to start is by using the word "write" and then working up to "say"—which are specific types of communication, until the *concept* of communication can be treated as a whole. This is what we mean by processing gradually or working with a gradient scale of processing.

The key is to always process with what *is* within the reach, accessibility and willingness of the Seeker as they are at present and then cumulatively extend that willingness to reach or *do*, building on the validation of what *is* within the known control of the individual. What we are then doing is reversing the programming and imprinting that has been stored, which has been found to limit the power of thought and consideration to smaller and more fixed

---

199 **Self-actualization** : bringing the full potential of the Human spirit into Reality; expressing full capabilities and creativeness of the *Alpha-Spirit*.

parameters. This degradation took place systematically throughout the course of the *Self's* own journey, projecting its POV through more and more rigid and condensed universes until finding all of the remaining "willingness" for consideration right now here in this beta-existence, tightly wound up as a box that we prize, guard and protect: the artificial personality.[200] It is *this* that entraps the Human Condition to beta-existence.

---

200 **personality (program)** : the total composite picture an individual "identifies" themselves with; the accumulated sum of material and mental mass by which an individual experiences as their timeline; a "beta-personality" is mainly attached to the identity of a particular physical body and the total sum of its own genetic memory in combination with the data stores and pictures maintained by the Alpha Spirit; a "true personality" is the Alpha Spirit as Self completely defragmented of all erroneous limitations and barriers to consideration, belief, manifestation and intention.

:: 3 ::

CONDENSATION OF UNIVERSES AND
FRAGMENTATION OF THE HUMAN CONDITION

Most sources alluding to the true Cosmic History—one that predates this version of planet Earth and even *this* version of the Physical Universe—are based on the most ancient writings we have access to; carefully scribed at the inception of writing systems during this current version of human civilization. These narratives generally relay a story using mythographic symbolism and as a result we find development of specific portrayals of a literal "mythology" that now, thousands of years later, have all been blown down for the "straw men" that they are. But!—they were all inspired by something, some memory, and the oldest of these recollections may be found on the *Arcane Tablets* and records from ancient Mesopotamia.

This is not to say that no other versions of human civilization took place on this Earth before the time of Mesopotamia, the Sumerians[201] and ancient Babylon—since we are now very certain that there is quite a bit more to this picture than we have ever been given as "common knowledge"[202]—however, as we worked through the possible natures of all previous potential civilizations, it became clear that there is an eradication of all recoverable written records every 10,000–12,000 years; which puts the physical archaeologist in a position of greater "guesswork" than they already find themselves in when attempting their own interpretation of physical remains.

All previous relays of "Cosmic History"—including the systemological interpretation of the *Enuma Eliš* that is provided in *"The Tablets of Destiny"* (*Liber-One*) and which directly contributed to the formation of our Standard Model—describe the linear pattern that is reflected in the "condensation of universes" and the degradation of the Alpha Spirit to become imprisoned within the POV and considerations of a hard-wired low-level Human Condition.

We know very succinctly *how* it happened; progressing from an Infinite Nothingness, to the individuation of the Alpha Spirit and onward to more restrictive and condensed manners of consideration and agreement about the Reality we exist within. This is reflected strongest in the "Gate" system of ancient Babylon; then afterward, in various lore of various *kabbalahs* and *chakras* that attempted to achieve the same results.

But the methods proposed over the past four millennium have done little for the Human Condition but enforce[203] more restrictive agreements upon its considerations and continue it

---

201 **Sumerian** : ancient civilization of *Sumer*, founded in Mesopotamia c. 5000 B.C.

202 **common knowledge (game theory)** : facts that all "players" know, and they know that all other "players" also know—such as the very structure of the "game" being played.

203 **enforcement** : the act of compelling or putting (effort) into force; to compel or impose obedience by force; to impress strongly with applications of stress to demand agreement or validation; the lowest-level of direct

on a path, that, by our measuring (using the Standard Model), is going to send the entire Physical Universe "out the bottom" in a relatively short time. The direction of movement, for too long, is always "downward" and sending our Awareness to follow along with these narratives has a tendency to lead our considerations to further "spiral inward"—validating and reinforcing the track we are presently on and its direction, should we continue to follow the standard issue programming. Without correction, this is really a very dangerous direction for an *eternal spirit* to be headed in. It undoubtedly serves as a basis for what has otherwise been buried in religious dogmas and cultural morality. But, since when did enforced agreement from the ignorant ever yield any progressive results?

As an individuation—an establishment of the personal identity continuum known as "I"—the Alpha Spirit first practices selectively being "out of communication" with what is "not-I" at the uppermost level of our Standard Model (*ZU-line*). It would be quite difficult to determine with pristine accuracy the relative quadrillions of years ago, by Earth-time standards, such a spiritual incident[204] would have occurred in the highest, most near-infinite, manifestation of existence as a "universe." Esoterically, it has been referred to as the "*I-not-I monad*," but it is the point of which the *Self* is "*Aware*" of its *Self* as *Self* at the highest point of *beingness* that is possible from this static[205] point.

In order to differentiate[206] "I" from the "not-I," the Alpha Spirit adopts a selective practice of imposing various barriers, communication lags and a distance to perceive across, so as not to simply be the mirror of the other "I's" or Alpha Spirits that also differentiate themselves as wave peaks or crests or the uppermost tips of the icebergs that emerge from out of the Infinity of Nothingness. *This much*, even the mystic and magician has had a certain inherent *sense of*, but the actually *knowing* and purpose behind the "spiritual visions" that come unbidden has not been properly discerned[207] in the past. There is a tendency to return all understanding to lower-grade reasoning and mythological systems as they might be personified crudely in "earth terms."

At the highest level of reasoning—prior to fragmentation of "Alpha Thought"—Self as Alpha Spirit was in a perfect command of its handling of "reality," able to shift its considerations

---

control by physical effort or threat of punishment; a low-level method of control in the absence of true communication

204 **imprinting incident** : the first or original event instance communicated and *emotionally encoded* onto an individual's "timeline" (their memory of events over the course of one or all lifetimes) that formed a permanent impression that is later used to mechanistically treat future contact on that channel; the first or original occurrence of some particular *facet* or mental image related to a certain type of *encoded response*, such as pain and discomfort, losses and victimization, and even the acts that we have taken against others along the timeline that caused them to also be *Imprinted*.

205 **static** : characterized by a fixed or stationary condition; having no apparent change, movement or fluctuation.

206 **differentiation** : an apparent difference between aspects or concepts.

207 **discernment** : to perceive, distinguish and/or differentiate experience into true knowledge.

and creations freely at will,[208] and without any automation being necessary. The whole of existence was consciously created and the Alpha Spirit *knew* it was creating the conditions of its own *beingness*; able and willing to do so to the furthest reaches of the cosmos[209] without inhibition.[210]

We speak in linear terms here, such as that the "Self *was* in a perfect command" or that the "whole of existence was consciously created"—and we often suppose about umpteen-zillion relative years that such spiritual events took place—but the irony here is that the static position of the *Self* as the Alpha Spirit continues to remain unchanged to this day; only its considerations and assumed POV has been affected—and this has taken place "willingly" through a successive and gradual decay of personal willingness, reach and ability!

Hence, as the only effective corrective measure that we have found able to be implemented, we simply run this programming and its circuits *backwards*, and are directing Seekers "back the way they came" and not imposing some new artificial course of action. The decay or pattern-cycle of time is merely relative to the consideration of events that have taken place. Back behind the considerations assumed and energetically imprinted on a personal timeline, the true Self—the Alpha Spirit—is still there, bright, beautiful and powerful; the "I" that *is* the *Awareness* of the individual... if you can but remember what you chose to forget...

Of course, the Alpha Spirit is not alone in the near-infinite "Creative Universe"—there is a sense that there are others; other Alpha Spirits that can also create. Although the Alpha Spirit is aware of the nature of their own creations and how they are separate as barriers and distances, and that these may be freely shifted and arranged at will, they *are still* barriers of a sort, even only to maintain ones own individuality. This is not a crime; it is completely natural. But, it also sets up the Self for a practice in being the effect of another being's creations. Since the Alpha Spirit can put up its own screens and images and walls, it can also shield or filter or create walls—around which the Spirit can be "surprised" by the unexpected creations found on the other side.

Rather than go into a lengthy linear narrative of Cosmic History at this stage of the *Grades*, it should suffice that we illustrate the very earliest basis for personal fragmentation with the simple individuation of an Alpha Spirit, its sense of "I" and its immediate progression to "I-AM" in contrast, protest[211] or contest with the "I-AM-*Nots*." At this uppermost point of static

---

208 **will** *or* **WILL** (5.0) : in *NexGen Systemology* (from the *Standard Model*)—the spiritual ability at (5.0) of an *Alpha Spirit* (7.0) to apply *intention* as "Cause" from a higher order of reasoning and consideration (6.0) than the thoughts found in *beta-existence,* where it manifests as "effect" below (4.0).

209 **Cosmos** : archaic term for the "physical universe"; implies that chaos was brought into order; includes past "universes" that we occupied.

210 **inhibited** : withheld, discouraged or repressed from some state.

211 **protest** : a communication objecting the enforcement or rejection of a prior communication; an effort the cancel the "is" quality, or existence of a previous creation or communication; the unwillingness to be the POV of effect.

beingness, the POV is unaltered, unable to be conditioned, impacted or harmed in any way. It is only later considerations and the degradation to a fixed and limited POV that incorrectly associates *Self* with other lower states and conditions of *beingness* whereby the Self "agrees" to be affected, simply as part of the game[212] taking place once entering the level of agreements found in these lower, more solid and condensed universes.

Willingness of reach and extent of withdrawal is learned and "tracked" along the personal timeline of an individual, carried from lifetime to lifetime and apparently only added to, and never discharged, subjecting the Alpha Spirit to accumulate more and more solid energetic masses as they descend into more and more solid, rigid and fixed POVs from which to grant *beingness* even to themselves.

When first the Alpha Spirit occupied POVs solely in what is now an "exterior" or Alpha universe existence, it may very well be that our original acceptance and rejection of energy communications was based purely on personal inclination and determination, but slowly these tendencies formed into patterns that we ascribed an "aesthetics" to—a sense of "beauty" or "ugliness" that transcends standard issue beta-concepts of analytical thought or even emotional imprinting. "Aesthetics" is actually not a quality that is inherent in beta-existence (this Physical Universe) and is applied from a higher state of *beingness* and *knowing* than can be measured as *beta-Awareness*[213] (meaning, of course, that it would be markedly higher than "4.0" on our Standard Model). Yet, these tendencies toward "aesthetics" laid further groundwork for potential miscommunication and automation.

It would be a steep gradient to send a *Grade-IV* Seeker back directly that "far" on their personal timeline using systematic processing and have it be effective. Of course, a highly actualized being would be able to fully change their considerations and willingness to reach by a matter of personal choice. That which inhibits executing this directive fully is not a fault with the *Self* or Alpha Spirit, but the accumulation of energetic fragmentation that exists along the lines between the "I" and command of its own *beingness*. Here, we are selectively illustrating the most accessible examples from our Cosmic History that will contribute to a workable understanding of the magnitude that we are dealing with. This is far from science-fiction; this is your spiritual beingness we are talking about.

What we find to be the case very early in our spiritual existence is no different than what we discover of the systems that we find our POV now occupying: the entire matter is related to a communication of energy along specific channels and circuits. The only thing that has

---

212 **game** : a strategic situation where the power of choice is employed or affected; a parameter or condition defined by purposes, freedoms and barriers (rules).

213 **beta (awareness)** : all consciousness activity ("*Awareness*") in the "Physical Universe" (KI) or else *beta-existence*; *Awareness* within the range of the *genetic-body*, including material thoughts, emotional responses and physical motors; personal *Awareness* of physical energy and physical matter moving through physical space and experienced as "time"; the *Awareness* held by *Self* that is restricted to a physical organic *Lifeform* or "*genetic vehicle*" in which it experiences causality in the *Physical Universe*.

changed is the rigid automation of these channels and the fixed solidity found in physical mediums of communication exercised in *beta-existence.* This is why an understanding of the systemology of communication, control and command is functionally useful and effective "across the boards" and not simply in one or a few specific instances.

So therefore, even in a near-infinite "Creative Universe" we can see seeds of fragmentation stirring. Just as we might withdraw our reach, close off communication and reject the creations of others, so too can others demonstrate a rejection of admiration toward our own creations. All various channels of energetic exchange are created and then treated with some consideration that could be very well reduced to whether or not we "*like*" such-and-such. And it has become increasingly apparent, as the *Self* moved its POV to more strict and narrow parameters of reality agreements, that the automatic nature of these inclinations and tendencies is not only a personal hindrance to Self-determinism, but it may also be manipulated and programmed, then passed off and accepted as a "personality" (though quite artificial in nature when compared to the truest ideal state of the Self as Alpha Spirit).

We have taken a peek back on the personal timeline to a point of origins for our individuated *beingness*, and this is where most narratives of the Cosmic History begin—though we have found it more workable, effective and practical for our purposes, if we focus specifically on the nature of *beta-fragmentation* in *this* version of the Physical Universe.

Once a Seeker is able to understand the base structure of their own programming here in this existence that the POV has been confined to, then we can press further into the higher frequency universes that we have "fallen" through to eventually arrive here. The way out is the way through; and we are charting our way back through the minefields and trappings that we have picked up along the way. No steps are to be skipped along this course if the footing is to be sure[214] and the way ahead is to be progressive. There is no going back to a point of ignorance now that we have witnessed the potential vistas that await the metahuman evolution of the Human Condition as *Homo Novus.*

Δ Δ Δ Δ Δ Δ Δ

Simultaneous with the condensation of universes and the agreement to confine the POV of *Self* to such universes, the *Self* became fragmented as a result of its own consideration that it could be. This started with the basic acceptance and rejection of energy very early on the timeline and later developed into various automatic mechanisms that we might consider "solids" from a purely physical perspective and semantic. As most of us know, "solid matter"—any solid form—in some way obscures the view of what is behind or in back of it. This is no less the case when we consider the creative forms and vices that were composed even in higher reaching spiritual universes.

---

214 **surefooted** : proceeding surely; not likely to stumble or fall.

Once we find the Self in a position of operating a Mind-Body connection from its Mind-System, the implanted programming for the Human Condition is actually a simple matter of logic—and entirely and systematically demonstrable with the Standard Model (*ZU-line*). The sequence of diminished ability and rigidity of fixed agreements and considerations becomes quite apparent when we take a step back and look at mechanisms inherent in beta-Awareness; but let us examine them one by one, from the inside out—keeping in mind that the *Self* had to be convinced to agree to each one of these conditions along the way as part and parcel for the course down the spiral. But hope is not lost. We are able to clearly see these systems for what they are now, and the way out is very much in reach.

Let us consider that at the root behind all systems in existence, the most fundamental "Prime Directive" is simply *to exist*. But, this is obviously not the only driving factor that may be installed in the Human Condition; it is simply the most basic one that applies to all existences. What we are interested in are the implanted directives that may be introduced to the programming of the Mind-System; which is essentially everything between "0.1" and "4.0" on our Standard Model—and which we have quite adequately divided into two "control centers" that relay communications toward the command of the Human Condition; one which is primarily "reactive" and one which is primarily "analytical"—referred to as the "Reactive Control Center" (RCC) and "Master Control Center" (MCC) at "2.0" and "4.0" respectively on our Standard Model.[*]

When we examine the lowest level fundamental base of the directive for *Self* "to exist" in *beta-existence*, and fixed to the POV of standard-issue[215] agreements to Human Condition, we find the conditional agreement that ties personal identity to any physical form—particularly a "human body" as a "genetic vehicle" for experiencing *this* Physical Universe. This is essentially the state of the standard-issue Human Condition as it exists today: fixed to the POV of *Self* as "aware" of sensing a "Physical Body."

Naturally, in our truest, highest, most ideal state, the Alpha Spirit cannot be affected; but *what if* the *Self* has made a sequence of agreements that led to an unknowing automated assumption that the "spirit" *is* the "body"? By forming this connection of pre-programmed considerations for a POV, we discover the "Mind-System" as it pertains to a "Mind-Body" connection. The POV of this Mind-System is not the actual *Self* either, but it *is* what many philosophers and mystics have treated as a "higher self" (from their perspective), although it is still very much tied to whatever implants and programs it is operating on, even in the upper-levels of *beta-thought* near "4.0" on our model.

---

[*] See "*The Tablets of Destiny*" (*Liber-One*) and "*Crystal Clear*" (*Liber-2B*) for an introduction to the Systemology of the RCC and MCC; both volumes are included in the complete *Grade-III* Master Edition Hardcover "*Systemology Handbook*" by Joshua Free.

215 **standard issue** : equally dispensed to all without consideration.

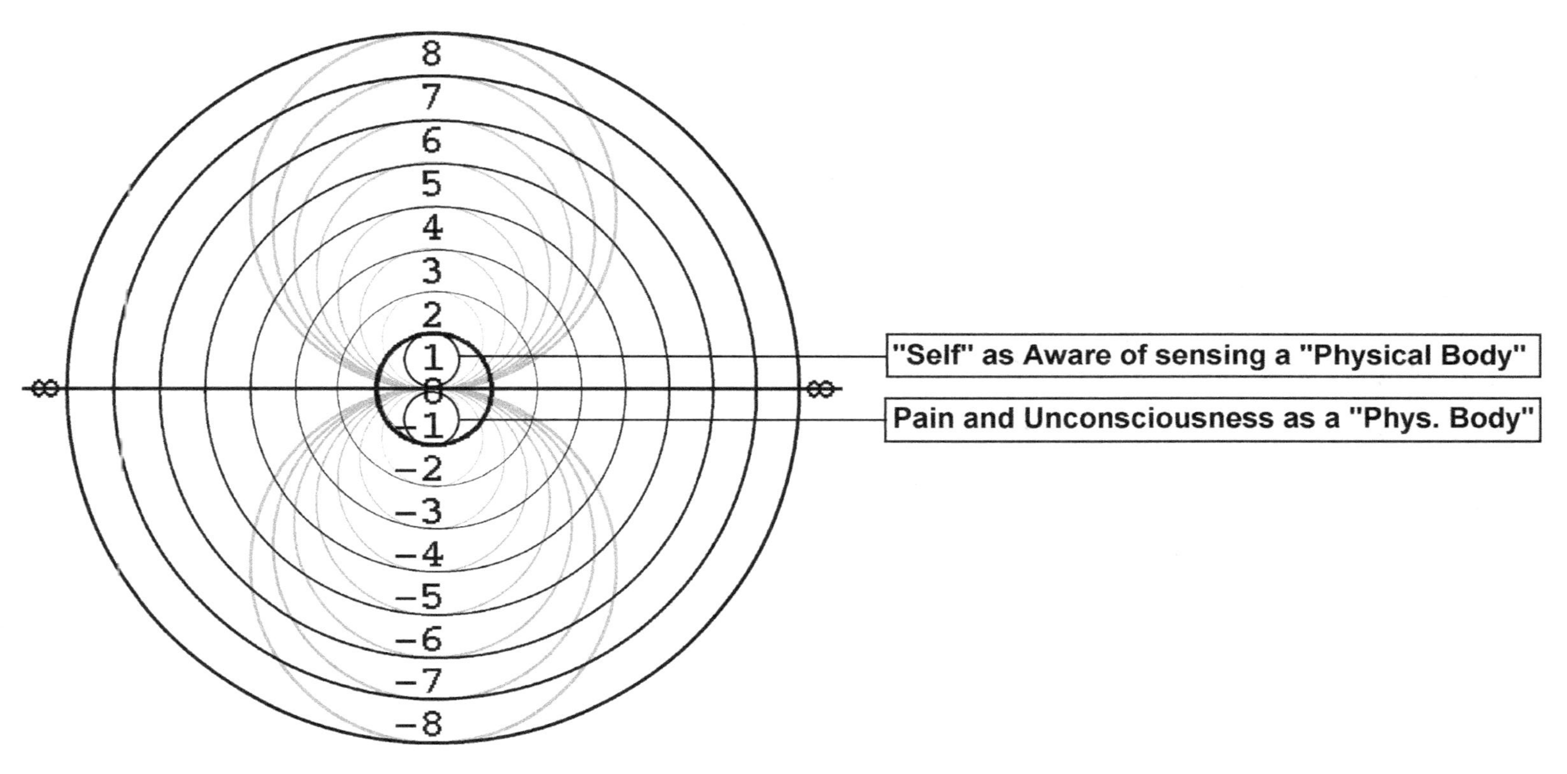
8
7
6
5
4
3
2
1
0
-1
-2
-3
-4
-5
-6
-7
-8
"Self" as Aware of sensing a "Physical Body"
Pain and Unconsciousness as a "Phys. Body"

Using the expanded version of the Standard Model (that includes sub-zero[216] classifications), it is easy to demonstrate the "fragmentation" inherent by the assumption of a "physical body" by simply threatening the *survival* or *existence* of that "body." It's literally that simple. Get an eternal being to believe that it is a mortal body and suddenly it can be conditioned with "pain" (from the body) and also additional programming through various states of "unconsciousness"[217] or reduced beta-Awareness. The manner in which this developed within and as the genetic organism[218] is not really very important—it may even have some kind of physical reasoning at a cellular level—but the facts remain quite visible to basic observation when we are dealing with this level of the Human Condition.

Sub-levels of the Mind-System, beginning with registry of "pain" and "unconsciousness"[219] as its fundamental encoding as a genetic vehicle, are deeply embedded and encoded implants; the nature of which has not been reached by modern sciences, philosophy or psychology.[‡] The primary issue, in the past, with referring to any of this as "un–" or "sub–" anything is that it has been presumed that these inner workings are fundamentally inactive, except during perhaps sleep or physical unconsciousness (such as coma states), and this is not the truth at all. If anything, it is the upper command from the Self (Alpha Spirit) and its own Mind-System (MCC) that periodically drops out, but so long as the POV is restricted to a standard Human Condition, these other sub-systems are *always on.*

This is very important for the Seeker that desires to access the upper-most potential of the Spirit; because while confining considerations of acceptance and rejection and reach and withdrawal to the POV of a genetic vehicle, the tendencies and automated inclinations would be as easily encoded as the mere infliction of "pain"—or really any sensation of discomfort—which can then be triggered later to direct reactive-response mechanisms outside to command and determinism of Self.

Do you now see why we emphasize the importance of *realizations* at this low-level occupation of POV that the "spirit" has succumbed to? Given that *this* is the current condition of the "human being" and its "Physical Universe" it is logical to presume that there is not much more room for us to go lower an existential universe. In fact, when this version of the

---

216 **sub-zones** : at ranges "below" which we are representing or which is readily observable for current purposes.

217 **biological unconsciousness** : the organism independent of the sentient *Awareness* of the *Self* to direct it; states induced by severe injury and anesthesia.

218 **organic** : as related to a physically living organism or carbon-based life form; energy-matter condensed into form as a focus or POV of Spiritual Life Energy (*ZU*) as it pertains to beta-existence of *this* Physical Universe (*KI*).

219 **unconscious** : a state when *Awareness* as *Self* is removed from the equation of *Life* experience, though it continues to be recorded in lower-level response mechanisms (fixed to a simulacrum or genetic vehicle) for later retrieval.

‡ Referring specifically to information given in "*The Power of Zu*" lecture series by Joshua Free, transcribed as a stand-alone volume and also included in "*The Systemology Handbook.*"

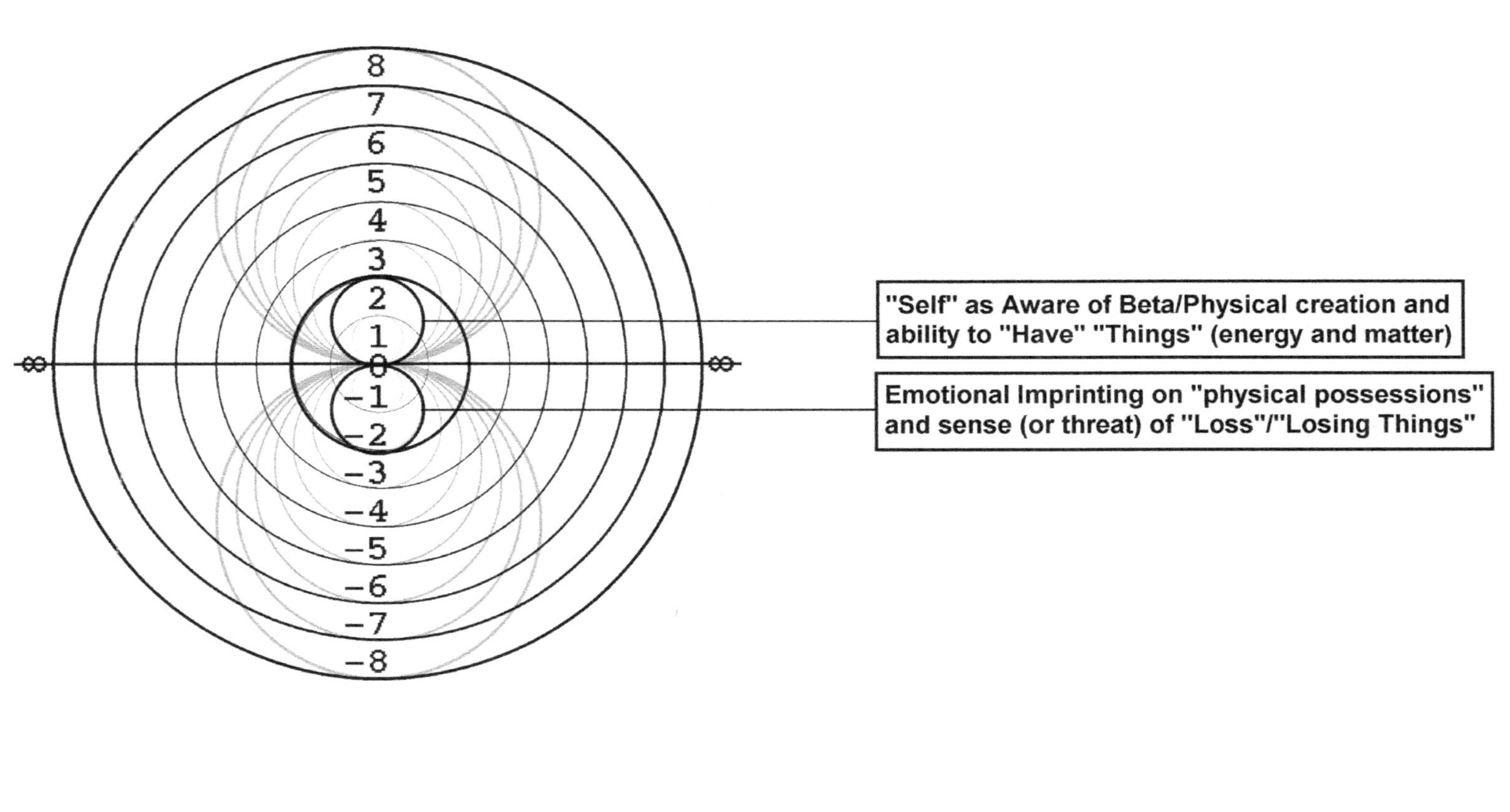
8
7
6
5
4
3
2
1
0
-1
-2
-3
-4
-5
-6
-7
-8
"Self" as Aware of Beta/Physical creation and ability to "Have" "Things" (energy and matter)
Emotional Imprinting on "physical possessions" and sense (or threat) of "Loss"/"Losing Things"

Physical Universe collapses in upon itself, the only POV left lower will leave *Self* "hanging on" as a mere enduring existence, functionally as static as a rock at a "zero-point" in juxtaposition to the true stasis of the Alpha Spirit.

When we moved past the first sphere on the Standard Model into the second, the other half of the systematic functionality of the "Reactive Control Center" (2.0) became crystal clear. Before the Self is "aware" of controlling a "body" it must *know* that it can *have* "things." And whatever it must *know* at this level of programming must be equal across all living systems that maintain an organic "reactive-response" nature. But again—the problem and solution were the same and it was not hard to realize that emotional encoding is linked to *having* and thereby could be affected by the opposite qualities of "loss." This was almost too easy to see and demonstrate systematically.

As a POV connected to sensory functions of an organic "genetic vehicle," the *Self* takes on the hard-wired programming that enables sensory reception[220] of data, using a "physical body" to perceive the material qualities of the physical environment. Programming and encoding at "2.0" on the Standard Model would thus be connected to a POV that is "aware" of the physical nature of creation (*beta-existence*) and that its composition of condensed solidified matter creates "things."

Unlike what we find with the capabilities of the Alpha Spirit operating within mental and spiritual universes—and where anything can be created and dispersed an infinite number of times—the creation and composition of energy-matter in this Physical Universe is primarily fixed and concentrated in such a way where "things" are unique from other "things." In the same sense that we discover an upper-level dichotomy[221] near Source of the "I-not-I," we discover that here in this Physical Universe, we have fragmented even our creations into "things" that are in exclusion to all other "things" and by this agreement we have the concept of "possession" or else the concept of *owning* or *having* "things." Since fluid creation of "things" is something of a scarcity as the universes condense, there is a heavier tendency to desire to "hold on" to things, because the Alpha Spirit has convinced itself it cannot just as easily create them again—any thing, mental image or sensation could just as easily be created again on one's own determinism and without reactivity.

Therefore, as soon as Self enters the POV of a game where material things are scare and therefore should be *had*, *protected* and treated with the same regard as we might to "mental" and "spiritual" things, emotional imprinting enters the picture—because now the Self can "lose" things; and this sense of "loss" (at "2.0") is treated as the same nature of threat to survival and existence as we might treat the care, protection and stewardship of a "physical body." Since we are programmed (at this level) that *knowing* and *being* is attached precisely

---

220 **receptacle** : a device or mechanism designed to contain and store a specific type of aspect or thing; a container meant to receive something.
221 **dichotomy** : a division into two parts, types or kinds.

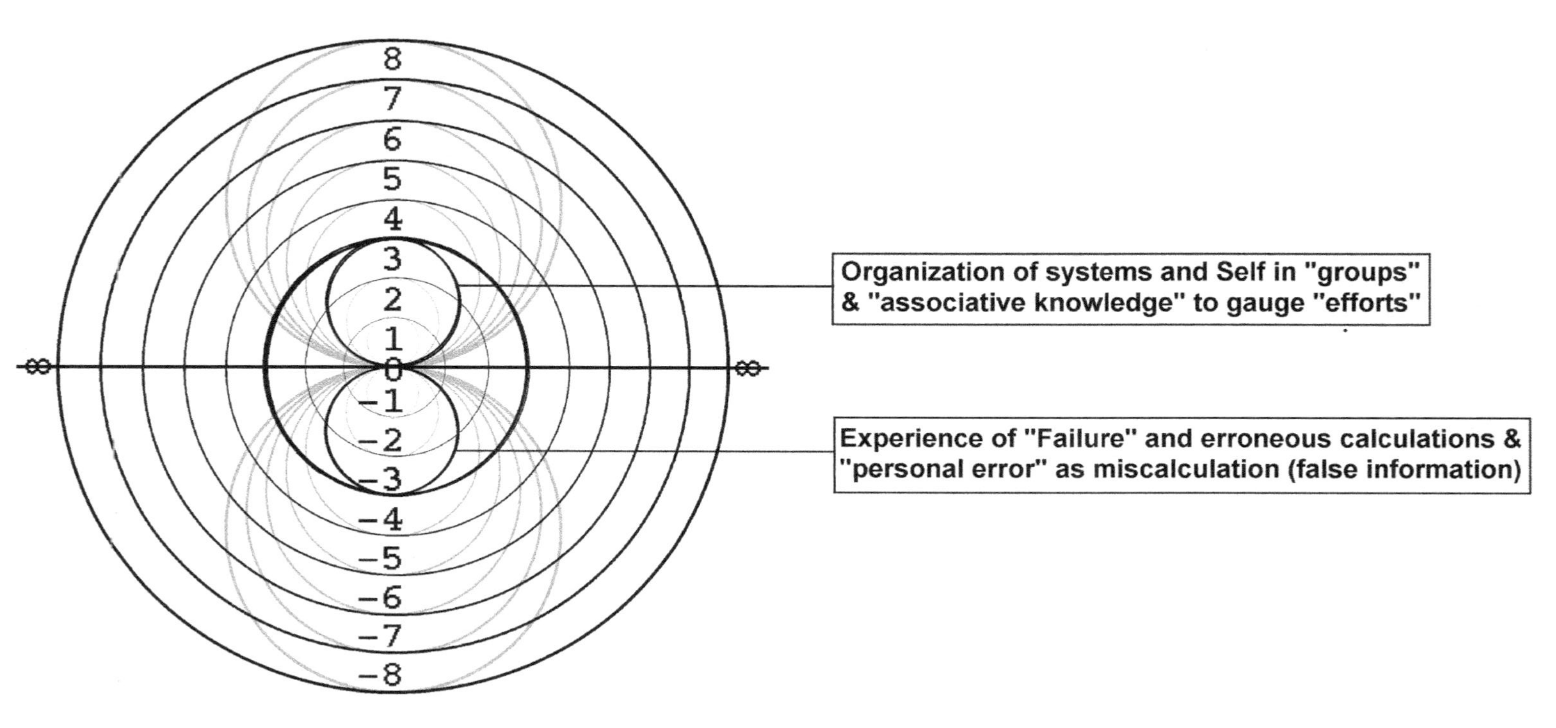

8
7
6
5
4
3
2
1
0
−1
−2
−3
−4
−5
−6
−7
−8
Organization of systems and Self in "groups" & "associative knowledge" to gauge "efforts"
Experience of "Failure" and erroneous calculations & "personal error" as miscalculation (false information)

with the *having* of things and the avoidance of *pain*, the fragmentation attached to the same is actually quite basic to work out—and much of this, a Seeker is set on the way toward with the former *Grade-III* material.

△ △ △ △ △ △ △

When treating the interior of the Mind-System maintained with the Mind-Body connection, the "RCC" constitutes only one part of the personal identity continuum (*ZU-line*). In fact, this part of beta-Awareness is mainly confined to one or another "genetic organisms." This means that the Mind can operate independent of the body, and a physical organism can function on basic survival programming with the Alpha Control[222] of the Mind eliminated from the equation.

To this upper part of mental command of beta-thought, we give the name "Master Control Center" (MCC) that is plotted at "4.0" on the Standard Model. Within this upper band[223] of "thought"—between "2.1" and "4.0"—the Self can occupy all manner of POV with the Mind-System that are independent to the location of a specific "physical body"; however, it has been found that these personal exercises of thought and imagination[224] are not Self-directed completely free of the "stuff" that the Mind-System has taken on when previously using the POV of "bodies." All of the "perceptions" and "data" are still stored as imprinting and programming encoded or embedded on the preexisting implants that we are now uncovering as we move through each *Gate* of *realization* and true *understanding*.

At the level of "1.0" we can say that the observable programming of the Human Condition from the POV of the genetic organism is "fight-versus-flight"—and this is compounded by the other half of the RCC system at "2.0" with a primary "stimulus-response" system. That about wraps it up for the "genetic vehicle" as a bio-chemical emotional entity.

Above this, yet also encompassing it whenever the "Mind-Body" connection is in place for *beta-existence*, we discover the larger framework of the Mind-System and its manner of computation is wholly different from the "identity-based" imprints of the RCC, which are

---

222 **alpha control center (ACC)** : the highest relay point of *Beingness* for an individuated *Alpha-Spirit, Self* or "I-AM"; in *NexGen Systemology*—a point of spiritual separation of ZU at (7.0) from the *Infinity of Nothingness* (8.0); the truest actualization of *Identity*; the highest *Self-directed* relay of *Alpha-Self* as an *Identity-Continuum*, operating in an *alpha-existence* (or "Spiritual Universe"–AN) to *determine* "Alpha Thought" (6.0) and WILL-*Intention* (5.0) *exterior* to the "Physical Universe"–(KI); the "wave-peak" of "I" emerging as individuated consciousness from *Infinity*.

223 **band** : a division or group; in *NexGen Systemology*—a division or set of frequencies on the ZU-line that are tuned closely together and referred to as a group.

224 **imagination** : the ability to create imagery in the mind at will and change or alter it as desired; the ability to create, change and dissolve mental images on command or as an act of will; to create a mental image or have associated imagery displayed (or "conjured") in the mind that may or may not be treated as real (or memory recall) and may or may not accurately duplicate objective reality.

found to generalize its "knowledge" as emotional encoding and includes all of the *facets* found within an experience indiscriminately.

To provide one example: the "concept of love" *could* be imprinted to a certain "identity," which could then be linked to a "loss" and finally associated with "pain." An individual would feel a reactive-response sensation of discomfort whenever the terminal of "love" is triggered. This surface layer of reasoning is actually built upon a much more complex array of associated thought, but it illustrates the point simply enough. But, to extend this point further in regards to personal fragmentation and programming: suppose an individual is hard-wired with an implanted personality that basically issues a command that "to exist is to love"?—and yet, when confronted by situations that communicate along that channel, other triggered imprinted encoding flashes up mental images or creates sensations as an automatic reaction, that may have first been developed to assist an organism in surviving by avoiding pain and loss, but now only fragments the clarity of any potential experience or knowledge so earned. And here we have systematically uncovered the root of *all* "human problems."

As soon as the Self is "aware" of, and operating *via* the Mind-System, those channels can be fragmented via implants and other programming. This operates along a similar pattern as what we see with the reaction-response mechanisms, except that the *facets* are not simply linked together as one identity. Within the band of beta-thought, or else the Mind-System that operates between "2.1" and "4.0" on our Standard Model (under the MCC), we find "grouping" and "categorization" of associated knowledge. Although analytical and not reactionary, the MCC can be fragmented with the experience of "failure" and "error." It uses its stores of information to calculate certain efforts and reasoning that connects the "mental universe" to the "physical universe" and then looks to the observable cues to see if this is correct. Misleading or erroneous[225] knowledge results in miscalculation and a deteriorated mental state through repeated invalidation[226] of personal data.

> Based on this systematization of knowledge, it is easy to see why the *third "Gate"* and *Grade* of our material is so heavily oriented toward the *"Pathway to Self-Honesty."*

It is this Self-Honest "clearing" of the channels that will restore the full command of the Mind-System to the Alpha Spirit. Too many of the functions and systems have taken on their own automation. The lower one looks on the Standard Model (*ZU-line*) of *beta-existence*, the more mechanistic and automatic the programs and circuitry seeks to become. These lower mechanisms tend to push an individual toward having a "pattern" or "tendency" and yet the more "other-determined" an individual experiences as Reality, the lower down the chain of *Actualized Awareness* they will be, until ultimately they will submit to agreeing to be the total effect of "forces" present in the Physical Universe. And, of course, once all sentient beings

---

225 **erroneous** : inaccurate; incorrect; containing error.
226 **invalidate** : decrease the level or degree or *agreement* as Reality.

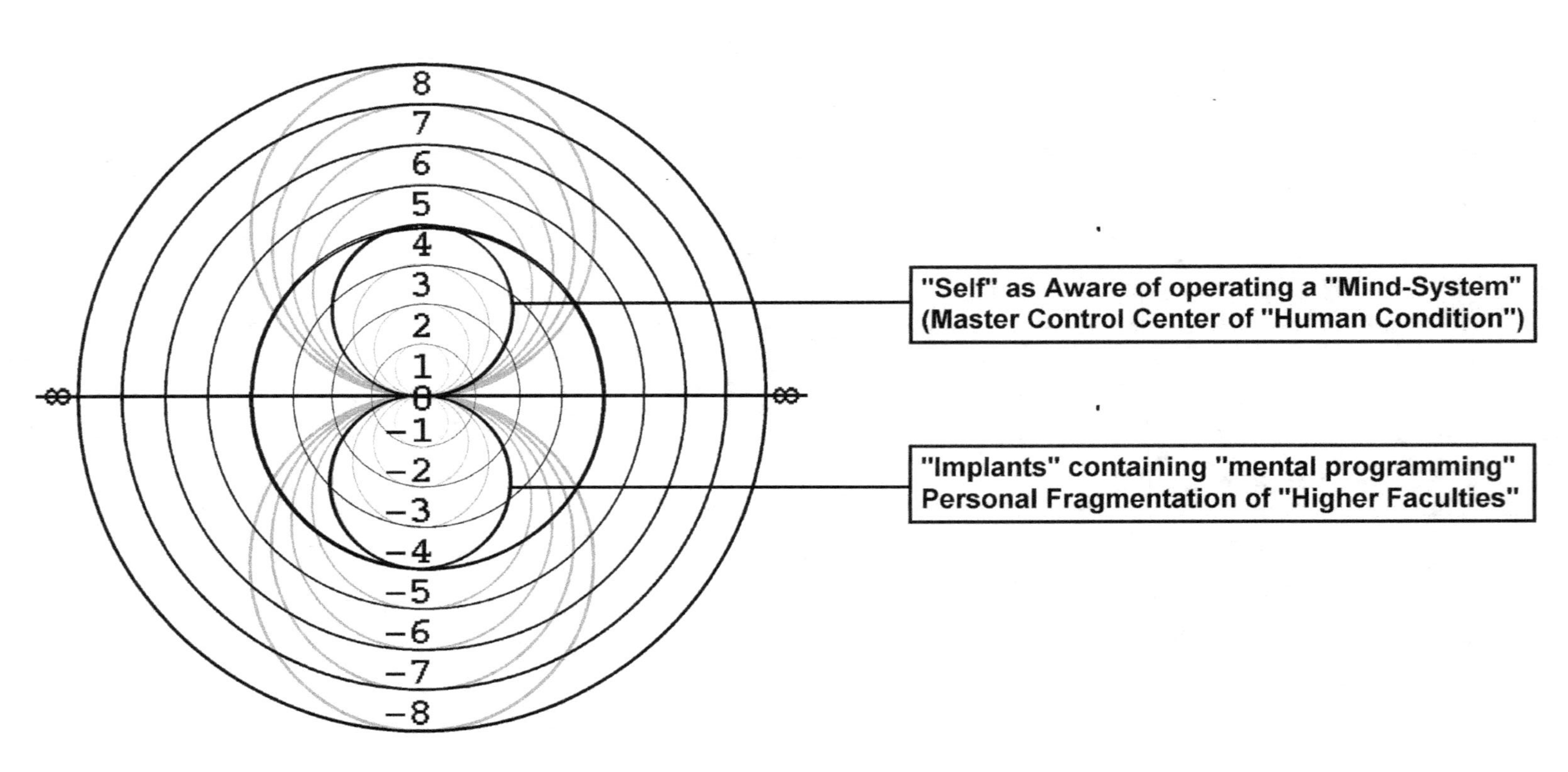

8
7
6
5
4
3
2
1
0
-1
-2
-3
-4
-5
-6
-7
-8
"Self" as Aware of operating a "Mind-System" (Master Control Center of "Human Condition")
"Implants" containing "mental programming" Personal Fragmentation of "Higher Faculties"

have agreed to become the total effect of one universe, there is no where else to go but down out the bottom.

The latest 21st century researches and discoveries of Mardukite Zuism and its Systemology have supported the fact that *there are solutions* to all of what is considered "human problems" and it has been further recognized that these problems and solutions are one and the same and given any consideration only because the Mind-System is running in such a way as to perceive them as such. The Mind-System is primarily occupied with computations of knowledge and effort in order to solve problems; and its computations are perfect to the degree of the programming that it is given and the state of fragmentation that it is operating at. To gauge this in any relative form, we developed the "Beta-Awareness Scale"[227]—as described fully in the text *"Crystal Clear"* (*Liber-2B*).[*]

Personal fragmentation in relation to systematic structure of the Standard Model of Beta-Existence is actually quite logical and follows an orderly sequence.

    4.0 Mind-System (Human Condition)
–4.0 ... able to experience "programming"

    3.0 Associated Knowledge (Calculations)
–3.0 ... able to experience "failure"

    2.0 "Having" (Emotional Association)
–2.0 ... able to experience "loss"

    1.0 Physical Body (Genetic Vehicle)
–1.0 ... able to experience "pain"

---

227 **assessment scale** : an official assignment of graded/gradient numeric values.
*   Also contained in the complete Grade-III anthology, *"Systemology Handbook."*

:: 4 ::

PROCESSING THE MIND-SYSTEM TO
DEFRAGMENT "PROBLEM" IMPLANTS

Our tradition of systematic processing is a highly specialized form of applied spiritual technology—a practical application of philosophy in such a way that seems to have alluded the reasoning and education of the human population. Methods of "home study" and "self-processing" are introduced to a Seeker in *Grade-III* materials—particularly *"The Tablets of Destiny"* (*Liber-One*) for fundamental instruction, with *"Crystal Clear"* (*Liber-2B*) supporting the aforementioned foundation with practical exercises.[*]

A Seeker may successfully employ *Grade-III* for *"Self-Processing"* and/or as a guideline for entry level *Piloted* procedure. A complete list of PCLs from *Grade-III* also appears in the appendix to *"Command of the Mind-Body Connection"* (*Liber-2D*), which supplements the basic *Pilot Training* introduced in *"Communication and Control of Energy & Power"* (*Liber-2C*). It is expected that a Seeker will have worked through these previous materials to complete their "Master Level" handling of this knowledge and technology before proceeding further in *Grade-IV*, which is presented to the Seeker and Pilot-in-Training as "Wizard Level-0."

A "Master" of any former esoteric or "mystery school" is expected to have achieved what was once universally identified as the "third degree" or *Third "Gate"* of potential realizations —a state that would put the initiate at the boundaries between physical and spiritual existence, with an alleged footing in either universe, but never fully in either. Rather than reaching further toward the Self-Honest clarity that lay hidden beyond just one more plateau of obscured thought, the path ended there. Oh, sure—the high-level initiates could take the continuity found at this higher vista and demonstrate to those that still worked below that there was a *"greater than"* (and even divide that realm into thirty-plus more "degrees"), but that was it. The computations devoted to associated knowledge allowed for near-infinite combinations, correspondences and "things to know" within the "mental universe"—but the Masters had not yet *gotten out* of their "heads" and were still operating very much *interior* to the fragmentation of the Human Condition.

Those individuals and Seekers that occupy their attentions on lower *Grade* work for extend periods of time begin to develop their personal universe around semantics and parameters presented within each potential paradigm. Even the physical scientist and mechanic runs in to this danger when linger too long, too closely, to the mechanized nature of the Physical Universe. The mystic magician and esoteric priest and priestess does the same when focused to strongly on the nature of "Cosmic Law"[228]—a Law which, if nothing else, only can be

---

[*] Materials from both *"Liber-One"* and *"Liber-2B"* are available within the complete *Grade-III* Master
    Edition hardcover anthology *"Systemology Handbook"* by Joshua Free.

228 **Cosmic Law** : the "Law" of Nature (or the Physical Universe); the "Law" governing cosmic ordering;

certain to dictate the course of energy and matter across the space[229] and time of *this* version of the Physical Universe. It could be thus used to create *another* version of this Physical Universe at a higher level of existence, but that is not necessarily the point. The point of that instruction was originally to remind the Spirit what it chose to forget; what it first chose to no longer be in communication with and then eventually obscured from view with automatically created barriers.

The truth being that we already came from a higher level of existence and only later continued to limit our considerations and the acceptance of enforced agreements to succumb to *this* one. Those *other* Universes are no less real than an abandoned property or ghost town that simply no longer receives creative attentions, as no POV is presently being occupied there.

As personal fragmentation and universal condensation became more refined—leading toward *this* Physical Universe—the *Self-determined* considerations and allocation of data regarding "conceptions" fell more and more in line with a particular "Law" or cosmic decree that is otherwise treated as some kind or another of "divine ordinance" or "Ordering" in religious explanations.

It is always presumed that this Order was "enforced" from the beginning—but there are spiritual indicators that this was a once equally created or agreed upon Order, which at the time seemed to "make sense" to all those involved; it seemed like a "good idea." But it was still a system—and as such, you will always find that certain individuals become the anomalies and kinks and trouble sources within the system; and along the timeline, it seemed to make good sense to the majority of those involved that there needed to be some kind of "penalty system" in place to keep this Order from going away, and if anything, make it more solid.

On the surface—"on paper" as it is said in this world—this graduation of Order into an "ethic" is a very logical progression of events, once multitudes of spirits started occupying "shared universes." The policing force of a higher universe would create a lower-level universe in which to "imprison" those that went against the established Order of a particular "Realm" or "Kingdom" (as they were sometimes known, as opposed to "universes"). But, given the decay of enough time and stricter and stricter system of enforcement, eventually everyone ends up in "prison" and it seems like the "universe" in the prison is now suddenly "where it's all happening." At some inevitable point, even the guards and upper-class citizens and kings all decide "it's the place to be." —And the whole mess starts up all over again at increasingly lower and lower levels. And now... *here* we are.

---

often called "Natural Law" in sciences and philosophies that attempt to codify or systematize it.

229 **space** : the viewpoint (or POV) extended out from any point out toward a dimension or dimensions; the consideration of a point or spot; the field of energy/matter mass created as a result of communication and control in action and measured as time (wave-length).

Keep in mind that in the very beginning, the Alpha Spirit was still more interested in their own personal universe and their own creations than the "game of universes." The real fragmentation of the individuated Self came later along with that, primarily as a result of interactions[230] with other energies and creations from other Alpha Spirits and the ensuing "experiences" that came to define an otherwise "artificial personality" laid over the original one of simply "I" as *Self.*

As Alpha Spirits began to experience creations of others, certain inclinations and tendencies formed. This is what we call "Alpha Fragmentation" because it took place on the timeline of the "spiritual universe" that is *exterior* to any *beta-existence.* It most certainly contributed to later programming, but in the spiritual existence we did not yet take on form that had to be fed or protected; and such low-level implant programming is dealt with more concretely in systematic *beta-defragmentation* processing. But, we should be aware, as we begin to weed out the fragmentation and "separate the wheat from the chaff" that we are dealing with matters that have very deep roots stemming from very early (relatively speaking) in our spiritual existence.

As we have impressed in our former *Grade-IV* material, the subject of "communication" is the common denominator of all personal fragmentation—and as a result it is where a Seeker begins. A basic methodology of systematic processing for this is found previously, so what we are concerned with in our present volume is taking the next steps in untangling the communication channels and circuits that are *interior* to the Mind-System; those that lead directly to the most fundamental "implanted problems" of the "Human Condition" as they are experienced in *beta-existence.* We treat this work as "Wizard Level-O" because it represents the bare minimum of actualization, as an upper *Grade* starting point, that we have found necessary for an individual to achieve effective success with any "higher level" work. While all of this material may make for an interesting read, if a Seeker is actually serious about their progress on the *Pathway to Self-Honesty* and beyond through the *Gateways to Infinity*, it is critical that they actual get the previous instruction and do the work, whether *Self-guided* or with the assistance of an accredited Systemology Pilot.

∆ ∆ ∆ ∆ ∆ ∆ ∆

As the Alpha Spirit (*Self*) became aware that there were other individuated Alpha Spirits (entities) that separated as an "I" from the Infinity, it became readily obvious that *Self* was not creating in a vacuum. Other POV were there to experience a creation; and the *Self* also maintained a personally determined POV to experience the creations of others. Interest and desirability became a personal taste regarding what was deemed acceptable and what was rejected. The *Self* discovered that it found things it did not like about the creations of others.

---

230 **energetic exchange** : communicated transmission of energetically encoded "information" between fields, forces or source-points that share some degree of interconnectivity; the event of "waves" acting upon each other like a force, flowing in regard to their proximity, range, frequency and amplitude.

Systematic Functions of the MCC

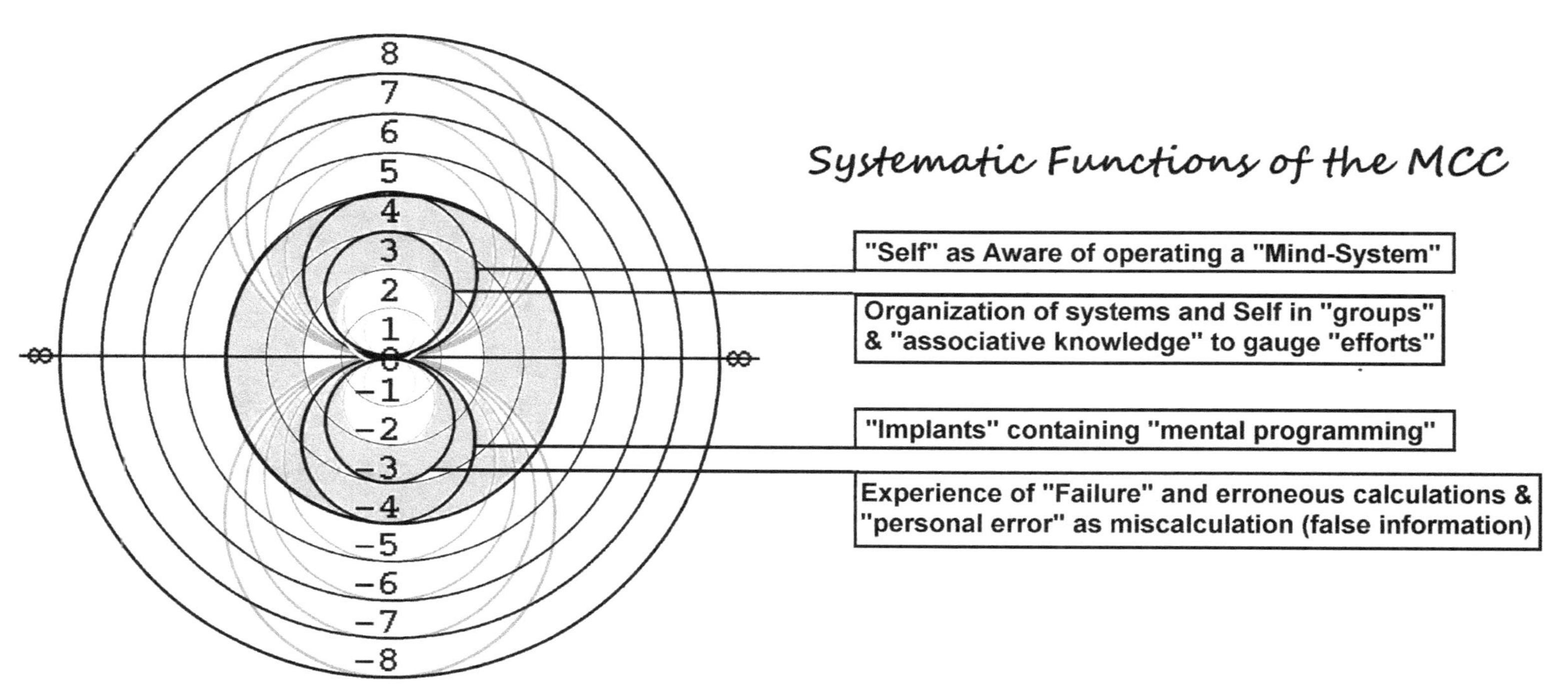

We do not find much fragmentation taking place with an individual and their own creations so long as they know they are creating it, can un-create it and re-create it at will. Such is the natural inherent creative ability of the Spirit.

Creative abilities of *Self*—or rather "considerations" of the same—only became fragmented as a result of other-determined acceptance and rejection of our creations in addition to the en-forced acceptance (or rejection) of other creations made by other entities. It is only here that we begin to see turbulence[231] on personal energetic communication lines. It is only here that we begin to see the Alpha Spirit demonstrate any "protest" against handling of communica-tion and creation in a universe. What this did, at *Alpha* levels of existence, was start creations in the direction of needing to be "more solid" in order to receive wider validation. "Protest" developed into an "insistence"[232] or else the repeated use of a communicated energy into a form that is simply more difficult to avoid acknowledgment[233] of or be ignored.

Handling of personal energetic communication does not, in itself, promote fragmentation. The Self is completely capable[234] of *Self-directing* any manner of communication and creation and accepting the same without fragmentation or spiritual turbulence. So, how does it hap-pen? It happens whenever anything starts to get set on automatic and the *Self* is no longer the determinant cause. This can happen very easily whenever there is an "insistence" be-cause, in the lack of an acknowledgment, the *Self* will continue to direct energy in a way that demands attentions and validation from others that something exists, is solid, as should re-main to be created. This patterned tendency may be found underneath "compulsions" to repeatedly perform an "intention"[235] or creative use of energy. In time, there is a lot of per-sonal energy directed by these "masses" (mechanism) unbeknownst to Self.

Systematic processing is intended to progressively return true inherent Self-determinism to an individual that has spent countless lifetimes developing further and further restrictions in their consideration and creative ability. If there is anything that "ages" or contributes to the decay of time for an "eternal spirit" it is *this*; and embedded beneath each and every one

---

231 **turbulence** : a quality or state of distortion or disturbance that creates irregularity of a flow or pattern; the quality or state of aberration on a line (such as ragged edges) or the emotional "turbulent feelings" attached to a particular flow or terminal node; a violent, haphazard or disharmonious commotion (such as in the ebb of gusts and lulls of wind action).

232 **insistence** : repeated use of a communicated energy into a form that demands acknowledgment, is more difficult to avoid or ignore.

233 **acknowledgment** : a response-communication establishing that an immediately former communication was properly received, duplicated and understood; the formal acceptance and/or recognition of a communication or presence.

234 **capable** : the actual capacity for potential ability.

235 **intention** : the directed application of Will; to intend (have "in Mind") or signify (give "significance" to) for or toward a particular purpose; in *NexGen Systemology* (from the *Standard Model*)—the spiritual activity at WILL (5.0) directed by an *Alpha Spirit* (7.0); the application of WILL as "Cause" from a higher order of Alpha Thought and consideration (6.0), which then may continue to relay communications as an "effect" in the universe.

of these imprints and compulsions is one more thread holding together a firmly agreed-to belief that *Self* and its source of energetic potential is in any way *finite*. There are some individuals that have already been hanging very high on the ladder and achieve great results and progress forward simply reading these general words and reminding themselves of their "purpose" and what brought them here. It is not, however, enough for us to treat few isolated cases—and instead have worked diligently on a standardized *Systemology* that could be applied equally to all Seekers at a particular given *Grade* of work.

We approach the subject of "Human Problems" as an extension of the communication processing introduced in *Grade-IV*. "Protest"—as a communication barrier—is a contributor to fragmentation when it becomes a "compulsive" mechanism, because any decrease in Self-determined ability is a decrease in *Actualized Awareness*. A Seeker is now directed to use the skills earned in their practice of our systemology to scan and identify any accessible mechanisms that may have been created, usually as "communication barriers," in order to maintain a POV of protest with a particular channel of communication. Once the Seeker is able to regain a *knowing* of their own creations, the command and control of these same becomes a simple matter of clear communication—and hence why these three "C's" are emphasized together in our material.

Many early implants were installed prior to the *Self* taking on a stringent Human Condition POV; they stem from deeply seeded imprinting that falls within the domain of "past lives." We cannot dismiss the obvious apparent nature of "past lives" and yet at *Grade-IV*, we are not directing Seekers (or their Pilots) to push this fact. There are many individuals not yet prepared to meet implications of "past lives" or their own spiritual timeline throughout "Cosmic History" and it is not suggested that this information should be in any way enforced until the Seeker is ready to confront it. Information contained in this manual is for convenience—and because our *Grade-IV* "Wizard Level-0" calls for its introduction in order to accomplish tasks of the *fourth "Gate"* and no more or less than this. If a Seeker were to track their successive "protests" and "rejections" back far enough on the spiritual timeline, they would undoubtedly reach these same conclusions—ones that can just as easily be demonstrated along the way with examples from the present lifetime.

As relayed in previous texts, energetic exchange[236] of communication is not a one-way flow when effective and primarily involves the interaction of at least two "terminals." This means anything we can treat as "communication" (from an energetic standpoint) can be systematically processed using *Grade-IV* methodology, involving defragmenting circuits along any communication channel using "Route-3."‡ If desired, the "A.T." extension of processing could

---

236 **energetic exchange** : communicated transmission of energetically encoded "information" between fields, forces or source-points that share some degree of interconnectivity; the event of "waves" acting upon each other like a force, flowing in regard to their proximity, range, frequency and amplitude.

‡ See the introduction course for *Grade-IV* contained in *"Communication and Control of Energy & Power"* (*Liber-2C*) and *"Command of the Mind-Body Connection"* (*Liber-2D*).

Systematic Functions of the RCC

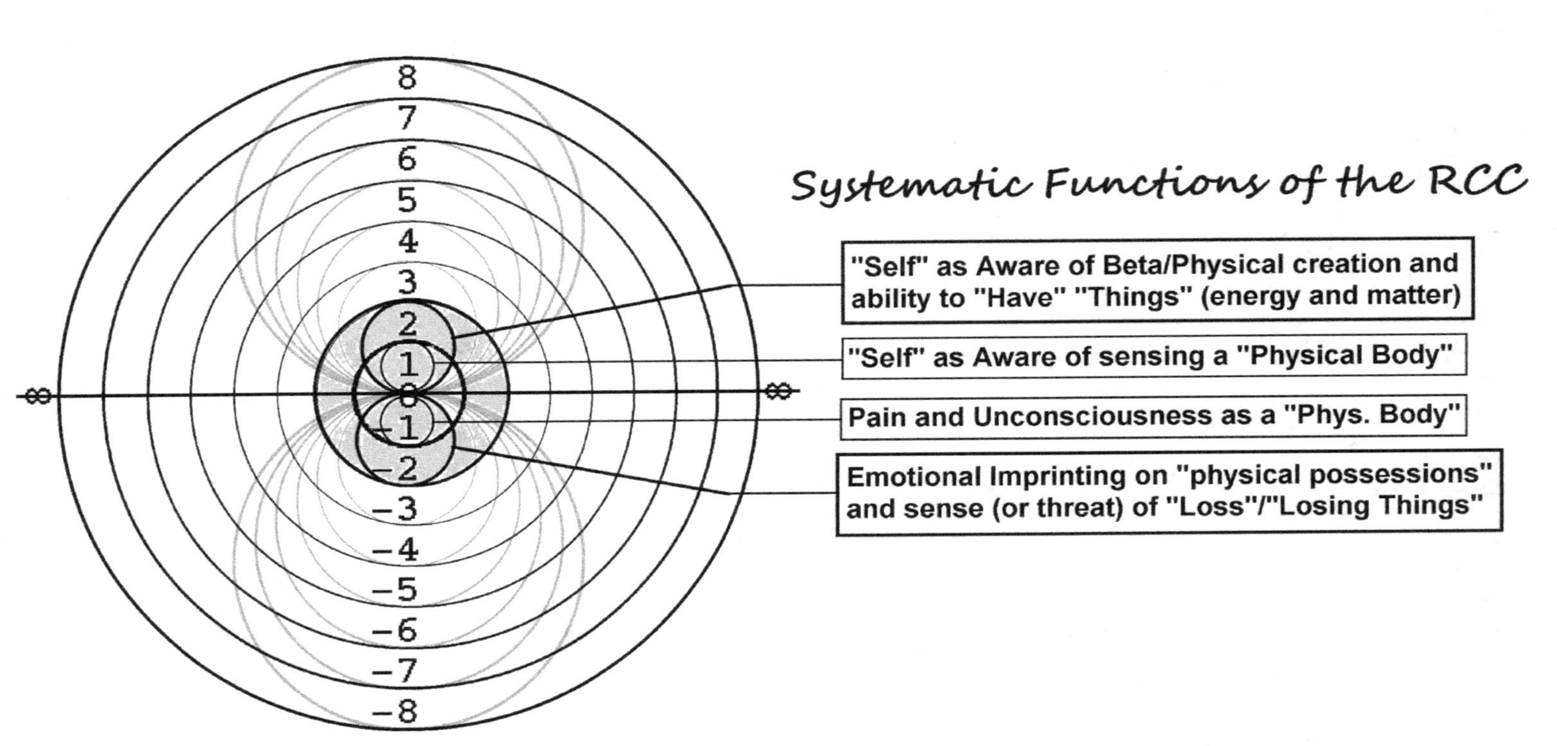

employ the skills of "creative mental imagery" that are emphasized later on in *Grade-IV,*[∞] including proper handling of "visualization,"[237] and the true creative abilities of the spirit, which have been long forgotten in contemporary "New Age" schools.

If we consider the POV from *Self*, the systematic processing to reduce the imprinted charge on channels that are presently engaged in "protest" and/or "rejection" (especially as an automatic tendency), would run repeatedly as as follows until a Seeker can resume control over the nature of the "communication barrier."

Circuit-1:	What is it that you are protesting?
Circuit-2:	How have you communicated that protest?
Circuit-3:	Who should be acknowledging your communication?
A.T. (*opt.*):	(*Visualize the terminal accepting the communication.*)

When a Seeker begins with their most recently and easily accessible examples, then processes them out successfully, it may very well be the case that an earlier similar "protest" or "rejection" appears from an earlier incident or event. This is normal and should be accounted for. This begins a personal journey of exploration into various degrees of higher and higher barriers that have been generated and agreed to over time. It did not happen all at once. Whatever state a Seeker is in today, they did not arrive there overnight. And the systematic imprinting and programming is laid in or keyed in differently for each individual along their course, although the encoding is embedded to a very similar line of implanted systems programming that is generally found across the board of all individuals. This is fortunate for us, because it gave us something to work with: a basis for systematically understanding and therefore resolving the subject of "Human Problems."

After the Seeker has worked effectively through the most accessible events regarding their own "protest" and "rejection" of communications from *other* terminals and source-points, the same systematic techniques can be used to approach circuits on each direction of flow regarding, in this case, terminals representing "protest" and "rejection."

For example, we can easily alter the above PCL format to apply to the POV of others: "What is (*terminal*) protesting/rejecting about you?" "How are they communicating?" "Who should acknowledge them?" and so on. This can even be applied to the third sphere of existence/influence regarding the mass communications of society as a whole, such as: "What are *others* protesting about *others*?" and so on. This actually has worked very well right now in presently reducing the reactive charge that is being experienced in the world today in

---

∞ Anticipated as "*Liber-3D.*"

237 **thought-experiment** : from the German, *Gedankenexperiment*; logical *considerations* or mental models used to concisely visualize consequences (cause-effect sequences) within the context of an imaginary or hypothetical scenario; using faculties of the Mind's Eye to *Imagine* things accurately with *considerations* that *have not* already been consciously experienced in *beta-existence*.

regard to all of the social disharmony. The time-tense of the PCL can be changed to "have" instead of "are" in order to better apply to past instances (once present ones have been reduced).

The *Self* does not only go out of communication with other "living" terminals or entities but also its own "things" and "creations." This is actually a very sad state of affairs for the Alpha Spirit, but it happens nonetheless and contributes to manifestation of greater pain and illness when identifying too closely with a physical body or genetic organism as itself. For example, when the individual experiences "pain" as tied to a specific part of the physical body or even a physical locale, the reactive-response tendency is to "avoid" such; which the RCC assumes its doing us a favor by manipulating our perception of reality to demonstrate the best chances of material survival. It simply registers everything that is connected to "pain" or "loss" as to be avoided, rejected and protested *automatically*, and therein lies the fragmentation.

A Seeker or Pilot can apply this basic systematic processing formula to defragment the channels that hold this type of communication all the way up through the *ZU-line*, applying the PCL as best fitting to each of the spheres of existence that we can reach to. "What have you protested about tht physical body?" "What have you protested about your home? Family?" "What have you protested about your career?" "What have you protested about your organization? Church? Neighborhood? City?" "What have you protested about the Human Condition? Life on Earth?" ...and so on.

Once a Seeker is brought to understand the nature of "protest" and "rejection" we then bring them to understand—and take back the command—of their ability to "accept" and "acknowledge" and then begin to go through the entire process again with an emphasis on times when they have failed to accept or acknowledge other terminals; times when other terminals have failed to accept or acknowledge, &tc. This can all be reasoned out very simply and an A.T.—"Actualized Technician"—or "Wizard" can effectively cap this off with the appropriate use of "mental imagery." We are simply working our way back out through the circuitry that has formed along one's own personal spiritual identity continuum. It requires some directed attention but is not impossible to process out.[238]

Δ Δ Δ Δ Δ Δ Δ

---

238 **"process-out"** or **"flatten a wave"** : to reduce *emotional encoding* of an *imprint* to zero; to dissolve a *wave-form* or *thought-formed* "solid" such as a "*belief*"; to completely run a *process* to its end, thereby *flattening* any previously "*collapsed-waves*" or *fragmentation* that is obstructing the *clear channel* of *Self-Awareness*; also referred to as "processing-out"; to discharge all previously held emotionally encoded imprinting or erroneous programming and beliefs that otherwise fix the free flow (wave) to a particular pattern, solid or concrete "*is*" form.

Our *Grade-IV* systemology is referred to as "Wizard Level-0" because it is encroaching on the levels of work that have only been dreamed about by mystical magicians and philosopher priests for thousands of years and are surpassing the theoretical structure we first built the methods upon by providing a new practical level of advancing through *Gateways to Infinity* that previously, from the standpoint of Master Grades, seemed fantastically out of reach and little more than spiritual pipe-dreams. It may be for this reason that so many other paths and traditions have fallen by the wayside before reaching this point; it may also be the case that after a few *had* reached this point, almost as if by accident, that they were not actualized enough to lead anyone else along, or else felt so possessive over the realizations inherent to the path that they would instead disguise and distort the knowledge so that those coming after would find only more fragmentation. The reasoning here is irrelevant now that the way ahead has been cleared. It is better not to become so occupied with such other trivial mysteries that we dedicate lives to little more than "conspiracy theorizing" and "ceaseless protest" against the systems of this Physical Universe.

Digging into the nature of "compulsive behavior" and "communication barriers" took several months of intensive research while applying *Grade-IV* techniques presented in former texts as "*Systemology Operating Procedure 2-C.*" All the while, we kept throwing the procedure at every *facet* of every *terminal* of "Human Problems" we could think of until eventually a pattern emerged that did actually correlate to our Standard Model. This led us to chart what we essentially considered the *interior blueprints* for "implanting problems in the Human Condition for beta-existence." After several revisions, this *Liber-3C* presentation of the outline of "Implanted Human Problems" seems to register quite effectively for all cases of the standard-issue[239] Human Condition.

    1.0 Physical Body (Genetic Vehicle)
        To Exist = Survive, Sustain, Eat, Sense
  –1.0 ... able to experience "pain" imprints

    2.0 "Having" (Emotional Association)
        To Exist = Reproduce, Protect, Satisfy, Cope
  –2.0 ... able to experience "loss" encoding

    3.0 Associated Knowledge (Calculations)
        To Exist = Organize, Reach, Cooperate, Compete
  –3.0 ... able to experience "failure" in evaluation

    4.0 Mind-System (Human Condition)
        To Exist = Solidity, Control, Division, Invalidation
  –4.0 ... able to experience "enforced agreement"

When and if automatic mechanisms, obsessions and compulsions are appropriately identified in a Seeker's case, they should scan and process only the most immediate imprinting events

---

239 **standard issue** : equally dispensed to all without consideration.

that seem to reinforce these tendencies and patterns. Trying too excessively to get at the true underlying core of implants will, every time without fail, force the Seeker to confront "past-lives" that they may not be ready to confront. It is for this reason that such knowledge and memory is often concealed (again, as a result of a "automatic mechanisms") until more readily available "solids" on that channel have been cleared away. Each time this occurs, the POV of the Seeker, as Actualized Awareness, is able to move further and further back in command of their own personal timeline, which is to say their personal identity continuum as reflected on the "ZU-line." The two are the same; one is merely a symbol of the other.

It is most likely that the recognizable mechanisms, obsessive fixations and compulsions that an individual exhibits are developed from increasingly growing energetic masses that form their own systematic structure as mechanisms. Wherever an implanted terminal node is in place within the *interior* of the Mind-System of the Human Condition, there is a channel of communication that should otherwise be under the command of *Self*. When these are allowed to go on automatic, then mechanisms form and energy is supplied to them from the Alpha Spirit but outside of conscious determination. There may have been a determined intent put in place at the start, but that has long since been overridden by allowing the automation to take over. This is one of the reasons that we impress the *Bell, Book & Candle*[240] techniques[*] of objective processing that return command and power of will and intention to the Self when conducting repetitive tasks.

Purposes behind systematic processing all surround freeing the *Self* by freeing up its "power of choice"—the fluid spiritual ability to "consider" and "create" without inhibition. When there is a tendency, pattern or compulsion that manifests in this lifetime and which appears to be outside of the Seeker's control, our current practice is to begin with *Analytical Recall* ("Route-2") and see if we can determine what the mechanism is a response to. It may require some repetitive inquiries before the "Mind" gives up its veil on the better answers; and it may very well be that these answers seem utterly ridiculous to the present lifetime or even *this* Physical Universe. Make note of them all; pay particular attention to any that seem to lead toward the type of *realizations* we are after in *Grade-IV* regarding the source of automatic or reactive-responsive tendencies.

---

240 **"bell, book & candle"** : three dissimilar objects that are kept accessible during a processing session (the book is often a copy of *The Systemology Handbook* or a hardcover copy of *The Tablets of Destiny* with the dust-jacket removed if it is less distracting that way); a term meant to indicate a Pilot's "objective processing kit" of objects generally present in the session room (accessible on a shelf, table or pedestal stands); in *NexGen Systemology,* the name of an objective processing philosophy pertaining to command of personal reality; historically, a formal ritual used by the Roman Catholic church to ceremonially declare an individual "guilty of the most heinous sins" as "excommunicated (to hold no further communications with) by anathema"—whereby a *bell* is rung, a *holy book* is closed and all *candles* are snuffed out—thus we therapeutically use the same symbolism historically representing religious fragmentation for modern systematic defragmentation purposes.

[*] See "*Command of the Mind-Body Connection*" (*Liber-2D*).

It may be too far of an entry level reach to simply ask a Seeker "What are you protesting with that behavior?" or "What were you protesting when you started doing such and such?"—but you may be able to approach the truth of things by applying the PCL: "What could you protest by doing *X*?" or something similar. Get the Seeker to come to some possible reasons why they would have set up that communication line in the first place. You may not be able to eliminate the original "Alpha" nature of the implant by processing events from *this* lifetime, but you *can* discharge some of the energy that keeps an individual from being able to ever reach back any further. Too much has been placed in the way—entire universes have been occulted from sight—and the progressive journey through the *Gates* cannot be bypassed in any way if it is to be true. No one can simply buy their way onward to the upper levels with goods wrought from *this* world and expect to arrive anywhere. Such organizations exist only to further entrap the *Self* into being an other-determined effect.

The tendencies toward "acceptance" or "rejection" are tied to closeness and proximity, or else the willingness to keep something close; meaning that it would be safe to keep something close. When we are dealing with tendencies and reactive-response mechanisms, keep in mind that these are all originally built upon a premise of logic that was structured for survival. That which was deemed "safe" or would contribute to survival was deemed "acceptable." This is all fine and good until it begins to have its values assigned (or "determined") to it on an automatic or reactive basis that eliminates the *Actualized Awareness* of *Self*. In the past we alternately have asked Seekers "What could you accept about *X*?" and "What could you reject about *X*?" until they realize that whatever they are rejecting, blocking, not wanting or avoiding is all based on past emotional imprinting. In other words, the "threat to survival" is not clear and present, but instead cast up into one's reality as a reactive-response with mental imagery that is intended to control thought and action in place of the Self-determined Alpha Spirit.

We have already made mention here of the position of the Alpha Spirit accepting and rejecting the creations of others—and having their own accepted and rejected—very early on in spiritual existence. It is not necessarily the intent of *Grade-IV* work to bring a Seeker back that far in their own personal processing; however, we introduce the information simultaneously within the "Wizard Grades" because the applications of the same technology is cumulative.

There is a "Route-3" PCL formula that may be used to weed out some of the fragmentation on these channels. After running the process on general events, the *term* "someone else" could be replaced with another appropriate "terminal" requiring specific attention. These PCLs are alternated within each circuit to prevent from getting "mentally spun" by running negative command lines. We extend the same principle from former "communication processing" and simply replace terminals and wording to apply to "presentation" of a creation as a more specific type of communication.

Circuit-1: What wouldn't (someone else) want you to present to them?
     What have you presented to (someone else)?

Circuit-2: What wouldn't you want (someone else) to present to you?
     What has (someone else) presented to you?

Circuit-3: What wouldn't (someone else) want others to present to them?
     What has (someone else) presented to others?

The same methods that are used to defragment the channels from events during *this* lifetime may be further applied later to access the deeper laden data of the implants that are carried between lifetimes. In some experiments that involved past lives, or the spiritual continuum, the first two circuits are reversed to see if any further information can be gleaned concerning a motivator. But, essentially, it is these same methods that were applied (simply at higher levels) in order to discover and verify at least enough certainty of the way ahead (at each stage) to formulate a solid presentation of a "current grade" for our Mardukite Zuism & Systemology books and organization.

:: 5 ::

PROCESSING TO HIGHER REALIZATIONS
IS ACCESSED WITH SELF-HONESTY

When we speak of the willingness to communicate, reach, accept, help and so forth, we are not enforcing any moral ethic on what an individual *should do*. What we have discovered is that the unwillingness on any lines is a point of other-determined control. An individual does not necessarily need to *agree* with anything, but it can be accepted as a communication and given an acknowledgment without actually participating[241] in its creation. We have found that we tend to add more solidity to that which we are rejecting and protesting than that which we simply allow to pass us by without interest. The ability to freely alter considerations and evaluations still allows the *Self* an ability to determine these on their own; but at least they *know* they are the ones doing it.

For optimum actualization, fluidity of consideration should be freely adjustable by Self. No, you don't have to agree with the way things are in the world, and you don't have to like so-and-so enough to want to be around them, and you don't have to necessarily become a "teacher" or "law enforcer" and you don't necessarily have to be a "vegetarian" or sit through long lessons about "math, science and history"... but shouldn't you be freely able to consider any of these things at will without some automated response-mechanism kicking in and taking command of your attention, focus, energy and behavior? You can practice this fluidity easily enough by repetitively[242] using the following formula with whatever you may find that you are protesting with a lot of emotional energy. This can also be practiced as objective processing on arbitrary items.

> —*Contact* the channel of mental energy connected with the "terminal" or "condition" that you are protesting.
> —Get the *Sense* (or POV) that <u>you</u> are "protesting" X.
> —Get the *Sense* (or POV) that <u>you</u> are "admiring" X.
> —Get the *Sense* (or POV) that <u>you</u> are "creating" X.

If this is not found to be effective after several runs through the process, it may be that the content of the channel needs to be handled on a gradient scale; meaning that perhaps too "large" or "general" of a terminal is being processed this way. Keep in mind that a Seeker can only reduce the emotional charge on *facets* and *imprints* related to events that they are prepared to "confront" as they actually are, not just as they seem to be from a fragmented state. This all has to be worked with—back and forth, alternating. The Mind-System appears to have a lot of checks and balances in place to make certain that this is the case; so that is what we are working with.

---

241 **participation** : being part of the action or affecting the result.
242 **repetitively** : to repeat "over and over" again; or else "repetition."

"Protest" is not the only relevant example of communication blocks and rejection of objective universes; it does however seem particular applicable to the present state of the world at the time of writing this manual. But again—this is not only relevant to now, but indicative simply of the point on a cycle that we have reached as a society, and as *this* version of the Physical Universe. An examination of human history will already demonstrate that this type of activity is cyclic in nature, but also that the results and products of each of these cycles is part of even larger cycles; systems within systems. This is illustrated quite clearly in our "*original thesis*" of systemology.[*] There are many visible signs that present conditions on this "prison planet" are on the cusp of great change; and it is of utmost importance for an individual to achieve *Self-Honesty* and arrive at higher *Gates of Realization* to avoid succumbing to entrapment further down into an even lower level universe than what where we have already fixed our POV.

It should be understood that when you are blocking, inhibiting, rejecting or protesting a line of communication, *you are doing* something. This is a projection of energy and active use of creative spiritual ability. This is far and beyond the simple "dislike" or "disinterest" or some creation, but specifically those creations that *Self* is actively (even if on automatic) projecting energy "against." This literally creates a spiritual "beam" of energy; a wave of personal emotional energy that is sent out *in response* to another waveform[243] so as to keep it at a distance. This, of course, also creates an additional layer to the individualized concept of "distance" within one's personal universe, and therefore "space."

Beyond word play we can literally suggest that anything you want to actively keep out of your "personal space" or "personal universe" is being *protested against.* We have realized that while this may have first started out as Self-determined and freely chosen, the patterned tendencies developed into their own automatic mechanisms using our own creative power. This too, may have first been the product of personal choice, but somehow we were also able to conceal these operations and their automation from our view, convincing ourselves that "out of sight is out of mind" when all of our experimental research concerning the Standard Model and the *interior* of the Mind-System would seem to suggest otherwise.

By taking over the responsibility[244] for the creative ability to be at every point-of-view, which is to say every source-point and effect-point of the creation, and yet not remain fixed or reactive to any (as demonstrated in the process that opens this chapter-lesson), the freedom to selectively apply these energies is returned to *Self.* Surely we must not be simultaneously

---

[*]   Available from the *Mardukite Zuism & Systemology Archives* and also reprinted in the complete *Grade-III* Master Edition anthology, "*The Systemology Handbook*" by Joshua Free.

243 **thought-wave** or **wave-form** : a proactive *Self-directed action* or reactive-response *action* of *consciousness*; the *process* of *thinking* as demonstrated in *wave-form*; the *activity* of *Awareness* within the range of *thought vibrations/frequencies* on the existential *Life-continuum* or *ZU-line*.

244 **responsibility** : the *ability* to *respond*; the extent of mobilizing *power* and *understanding* an individual maintains as *Awareness* to enact *change*; the proactive ability to *Self-direct* and make decisions independent of an outside authority.

creating, accepting/admiring and protesting a specific "concept" all at once, but at this stage of the game, and given how many implants and lifetimes of programming are in place, can we be absolutely certain of that on an energetic level? Well, the answer is "yes," and the proof in that comes from putting the actualized POV of *Self* into any of the positions at will, running it to a certainty that *"you"* are doing it and then freely able to do some other part or not at all; confronting without reactivity. We have applied the same technology to systematic processing of "emotional reactivity" in *Grade-III*.

When we approach energetic qualities of universes from an even *higher* level of reasoning that is currently in development for the *A.T. Wizard Grades,*[‡] it may very well be the case that our automatic "rejection" and "protest" of *this* version of the Physical Universe is what fixes our POV to these *beta-existence* "anchor" points[245] and which holds the material existence around us in place and all the while actively making it more solid. This may seem incredibly "metaphysical" for a *Grade-IV* Seeker, but it is something to consider since we discovered that whatever we are willing to take the full responsibility for creating no longer seems to have an effect on us.

Δ Δ Δ Δ Δ Δ Δ

An individual declines on the scale of *Actualized Awareness* as their considerations become more solid to meet that of the Physical Universe. As the perceived fixed solidity of the Physical Universe increases an individual finds that they are the *effect* of more and more low-level "problems." This fixation is what reduces the *Awareness* necessary to confront any greater level of problem solving. An ability to approach the problems and consequences (or penalties) increases as a Seeker reaches higher vistas of understanding and the determinism to shift around the considerations of "what is a problem." The Physical Universe therefore is able to maintain a hold on the POV of an individual to the same degree that the individual permits a fixation on the "mystery" of solving the implanted problems of the Physical Universe. And as such, the decision-making and actions employed by the Alpha Spirit is treated as *Game*, which is a concept quite prevalent[246] within our systemology.[*]

---

‡   Systemology *Grade IV, V, VI* and *VIII* are our "Wizard Levels" 0 through 3.

245 **anchor (*conceptual*)** : a stable point in space; a fixed point used to hold or stabilize a spatial existence of other points; a spatial point that fixes the parameters of dimensional orientation, such as the corner-points of a solid object in relation to other points in space; in *NexGen Systemology*, "beta-anchored" is an expression used to describe the fixed orientation of a viewpoint from Self in relation to all possible spatial points in *beta-existence* ("physical universe"), or else the existential points that fix the operation of the "body" within the space-time of *beta-existence*.

246 **prevalent** : of wide extent; an extensive or largely accepted aspect or current state.

*   The former academic research regarding "systems theory" and "games theory" is treated more directly at the higher Wizard levels/grades, but the concepts have been an underlying integral to our Systemology since its inception.

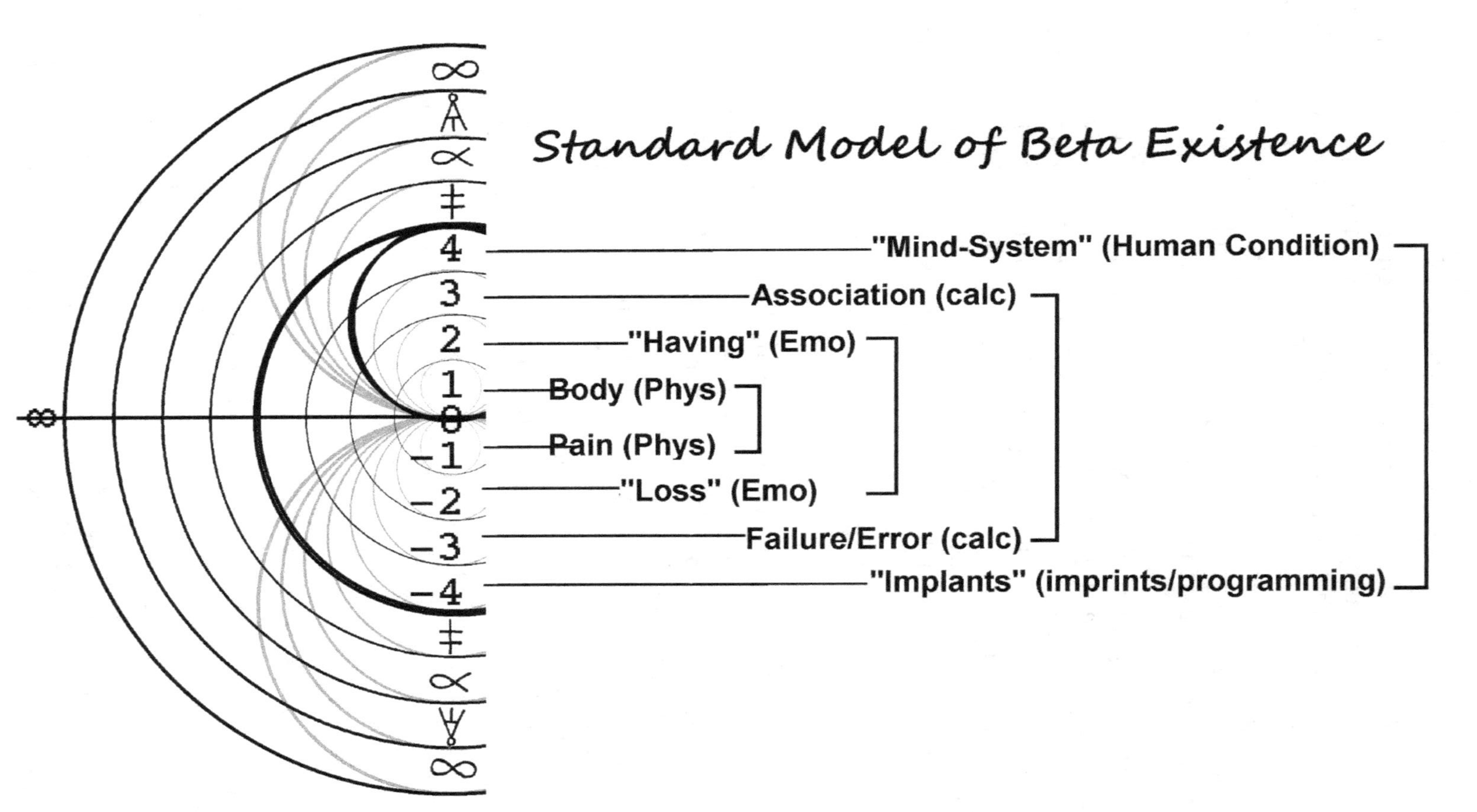

Standard Model of Beta Existence
"Mind-System" (Human Condition)
Association (calc)
"Having" (Emo)
Body (Phys)
Pain (Phys)
"Loss" (Emo)
Failure/Error (calc)
"Implants" (imprints/programming)

An understanding of Human Problems, as general given to the masses, is reduced to the level of the Physical Universe that "two plus two" is a problem requiring resolution. For many, the very idea of an "equation" denotes a problem to solve. But the problems implanted in the Human Condition are designed to be more complex than one-sided equations. The Mind-System has no issues solve material equations it can find interesting. The *real* problems that face the Human Condition are those that "*must* be solved; but *cannot* be solved" and therefore get a person hung up on a "mystery" or "unknown"—which is to say a perpetual "maybe." So long as a "problem" is suspended in such a state, so too is the *Awareness* of the individual.

A *true* problem involves "conflict"[247]—whether of a conceptual thought or an intention to *do.* For example, an individual has thought put out on a line and it comes up against a counter-thought coming in from the other direction and there you end up with this cluster mess of energy that we can call a *real* problem. A source-point or individual is projecting a concept of "is" or "am" and they are coming up against the "isn't" or "not" along the same channel.

This same principle applies to the will to intend to *do* something, or direct an action or motion of energy, which we refer to as *intention.* This is much different in practice from simply deciding about how one is going to solve an equation. Imagine being told you needed to take a test and you agree to locate yourself in space-time to take this test, but you were not given instructions on the material being tested or even where the test was to be taken. This absence of information poses a serious problem, and yet the directive is now in place that you *must.* Or perhaps you have the intention of focusing on taking a test and the person next to you has an intention of distracting your attention. Anywhere we find two persisting flows of energy in direct conflict with one another, we have a "problem."

We discover something of problems as we experience consequences (or penalties) involved with making choices in any "game-style" universe—which is the basic format of most objective realities that have since solidified. In the example of *games,* the counter-intention may be a barrier that is in place or it may be the opposition created by another player.[248]

Very seldom does an Alpha Spirit find itself having *real* "problems" when there is nothing in conflict, because the energy will continue along its directed course until meeting a condition that acts as a barrier, which in the case of "problems" we tend to treat in regards to "con-flict." This can also manifest and imprint similar to the way that persistent protest develops into compulsions; the fixation on the conflict can develop into obsessive automatic behavior mechanisms when left unchecked. The "problem" does not go away, so the individual sets up

---

247 **conflict** : the opposition of two forces of similar magnitude along the same channel or competing for the same terminal; the inability to duplicate another POV; a thought, intention or communication that is met with an opposing counter-thought or counter-intention that generates an energetic cluster.
248 **player (game theory)** : an individual that is making decisions in a game and/or is affected by decisions others are making in the game, especially if those other-determined decisions now affect the possible choices available.

a reactive mechanism to just keep pounding away on it indefinitely. This begins to filter and affect the way in which all "problems" (and even the ability to responsibly "confront") are handled.

By incorporating a dynamic POV of Self, we can understand semantics of "one-sided" versus "two-sided" problems quite reasonably. When two intentions are countering each other on same channel, the individual hits it as a problem. It only becomes a *real* problem if they get stuck on it. And problems have a tendency to cluster and accumulate when they are not managed properly; and this excess creates confusion. In an ideal well-adjusted an actualized state, an individual should be able to adjust their POV freely between the two aspects of *conflict* and simply resolve one or the other by some other avenue or application of energy. Once one or the other sides is no longer in conflict, the problem dissipates in that there is no longer a suspension of energy forming some kind of "obscuring mass" or "fragmentation." Often the solution is simply finding alternative considerations for one side or the other.

The standard issue Human Condition becomes easily fixated on "trying to solve" very low-level problems, but problems are simply a part of the *games* of this universe and in many ways become underlying implanted "purposes" that drive one or another "personality" sets. Even taking on a particular "phase" (or "identity package") tends to come with its own set of built-in "problems." But the Alpha Spirit enjoys playing games and only becomes fragmented by them under the consideration that the *Self* and the POV of the "phase" or "personality" assumed (or implanted) are the same. In this case, the Alpha Spirit begins to lose its own freedom of ability and level of *Actualized Awareness* by increasingly agreeing to make the reality of the POV more solid, thus fixing attentions and conceptual considerations to the same.

Problem clusters turn into confusion when too many of the intersecting or conflicting energies are not *Self-determined.* When left to automation, the artificial personality package will engage with its own preset means of treating the "problem" based on its *facets,* and this we know from observing the behavior of the RCC, which imprints all *facets* of knowledge toward "pain" or "loss." The problems continue to exist because the individual is still providing the active energy for it to exist; and for many, this is what provides a sense of "purpose" in life, because life in this universe is wired for solving problems. So, the implanted problems are simply agreed to and accepted as part of this game in order to have an activity here rather than invent and play a new better game.

When a two-sided problem (intention/counter-intention; purpose/counter-purpose, &tc.) is locked into a "cluster" of confusing energy, the individual is likely unwilling to confront (face up to) one side or another of the problem. The other side of these problems does not necessarily have to be initiated by another individual, but simply by another opposing terminal that is introduced as an arbitrary or variable, such as a "rule" of the game. One side of the problem may be an implant on the individual whereas what it is coming up against is a barrier placed in the game. Combined, the two are supposed to provide some sense of purpose in the game, matching freedoms and purposes against the rules and barriers. And so the world continues on.

For example, one common implant for beta-existence states that: "to exist is to be in a body which must be protected to endure." So we solve this with the concept of houses. Now the problem becomes that we "need a house to live in to survive, but houses also cost monies." An individual can have a lot of emotional imprinting and mental fragmentation regarding the concept of money. So, now the game for the Alpha Spirit has been reduced to resolving the "financials" to resolve sheltering a body that it has been tricked into agreeing that is Self and must be protected in order for Self to survive. Is it any wonder that the god-like Alpha Spirit is waking up to the fact that it has slid into a very low-level of considerations just to keep playing a game with everyone. But here the Seeker should see the difference between the "imprinted directives" and the "barriers"—both of which are still subject to an incredible amount of fragmentation.

Now again, ideally, an actualized person should be able to manage their considerations of a problem without getting all bungled up, but past fragmentation (imprinted on the beta-personality of an individual) can play a strong part in affecting how the individual perceives and confronts the external universes. For example, an individual should be able to decide on the appropriate type of home for their means and supply the appropriate means to it by operating in a Self-determined manner; but of course there are so many *facets* of the Human Condition that are likely to get in the way for a fragmented individual.

Δ Δ Δ Δ Δ Δ Δ

The *real* problem that a person cannot get beyond is the one that they cannot take responsibility for on either side. This means they are unwilling to appropriately *be* on either side, which means they are unable to properly confront it. We use the term "POV" or "point-of-view" to denote the sense of *beingness* that can be acknowledged in all things. This is one of those important truths that leads us to consider occupying more than simply the POV of the genetic vehicle and its artificial personality package—but also to recognize when we have adopted another person's POV unknowingly and are really operating from their programming, for whatever the reasons the RCC has deemed it more effective for material survival to do so (and keep in mind that the RCC does not operate on rationale and reason).[249]

The nature of Human Problems is generally cyclic and continuous when left to standard-issue conventions. As soon as a fragmented individual begins to think clearly on the matter

---

249 **rationality / reasoning (game theory)** : the extent to which a player seeks to play (make decisions, &tc.) in order to maximize the gains (or else survival) achievable within any given game conditions; the ability and willingness of an individual to reach toward conditions that promote the highest level of survival and existence and make the best choices and moves to see the desired goal manifest.

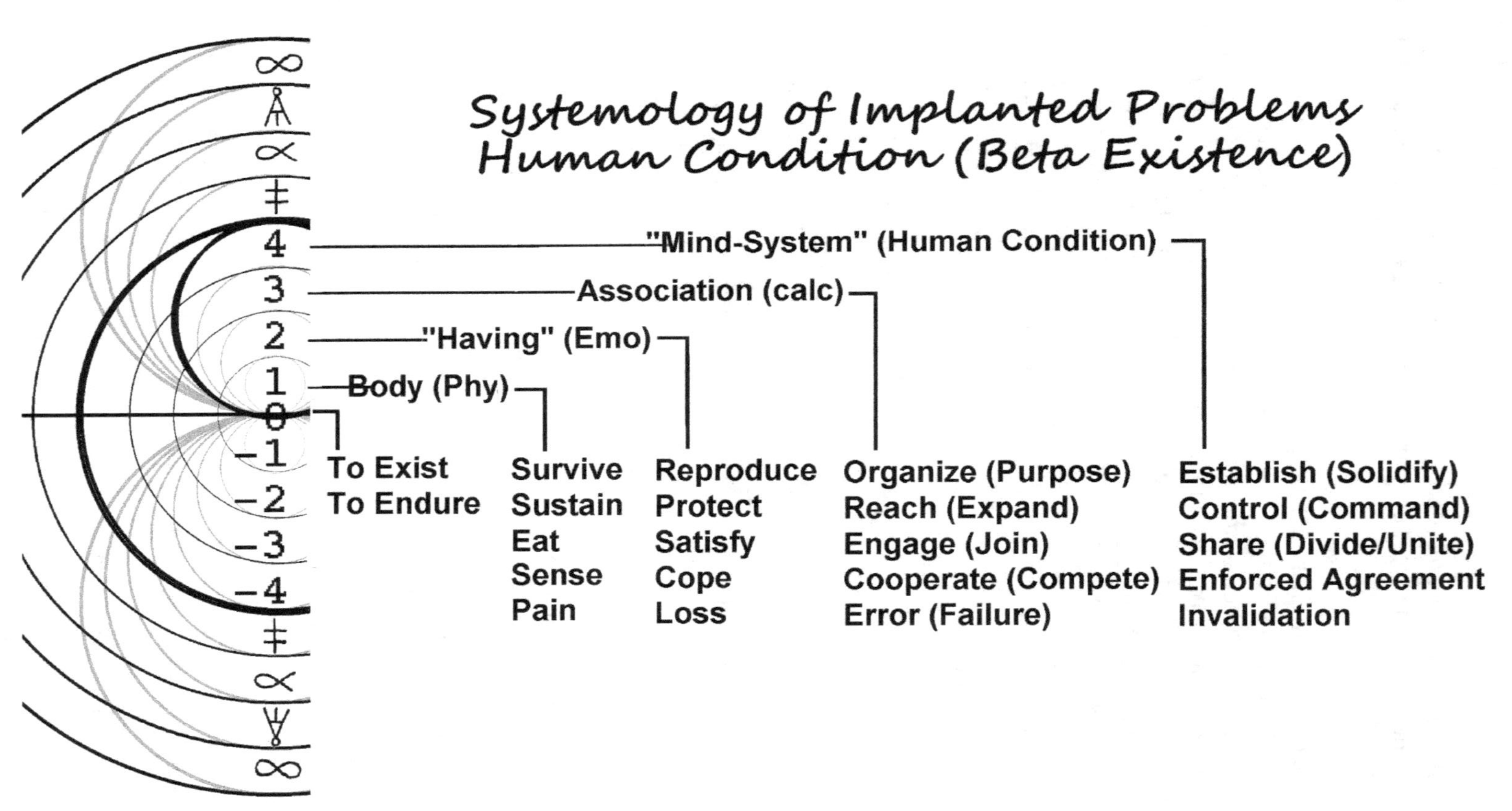

Systemology of Implanted Problems
Human Condition (Beta Existence)

"Mind-System" (Human Condition)
Association (calc)
"Having" (Emo)
Body (Phy)

∞
4
3
2
1
0
-1
-2
-3
-4
∞

To Exist
To Endure

Survive
Sustain
Eat
Sense
Pain

Reproduce
Protect
Satisfy
Cope
Loss

Organize (Purpose)
Reach (Expand)
Engage (Join)
Cooperate (Compete)
Error (Failure)

Establish (Solidify)
Control (Command)
Share (Divide/Unite)
Enforced Agreement
Invalidation

of shelter, the imprinting and encoding on the subject of money gets them all confounded. Then as soon as they start to apply some attention to the money half, that fragmentation stirs up so many emotions that now they can't think clearly about shelter needs. They have fallen into a trap that honestly very few in today's society successfully rise above without critical third-party assistance. Even then, that may only solve the immediate proximity of a problem that the individual is still no better equipped to manage properly in the future without systematic defragmentation of these channels.

One of the main issues, long observed on the Human Condition, and defying all reason and logic for the social sciences: when an individual has found themselves stuck in this loop with an unwillingness to confront it, personal energy goes into reinforcing the solidity of both sides of the problem—putting out energy, and putting out an extension of more energy against it. In essence, the *Self* is placing its POV between the opposing forces and holding both of them in place, hence maintaining the confusion by an unwillingness to *change.*

Self will only knowingly give up its hold or fixation on something if it can be demonstrated that it can resume this creation or communication on this channel at will. In systematic processing we demonstrate that any condition can be created again. This is as simple as getting a Seeker to invent problems. Once they do this knowingly and willingly then control over the management of problems will return and a deeper truth will be revealed: that although barriers to the true creative ability of the Alpha Spirit have been blocked on the descent through a condensation of universes, in order to balance the Infinite of Nothingness with an infinite potentiality of form, the Self still wants something to *do*, and it will set up automatic mechanisms to create those conditions if necessary. If a Seeker can practice being at cause with the creation of problems (as an imagination exercise) there is less likely a chance they will be compelled to compulsively do so on an automatic basis.

Another type of problem also operates within the Human Condition, called the "hidden implanted standard"—and there are several of these, most of which are formed during the Alpha Defragmentation process, meaning even prior to Self ever even assuming a humanoid POV in this Physical Universe for the first time. There is also a "hidden implanted standard" for beta-existence, and this is the underlying "directive" that the artificial personality is composed around. It is the ultimate filter or piece of fragmentation by which everything that is experienced or directed in the Physical Universe must first pass through. It may even be found in the patterned responses to PCL by a perceptive Pilot—or it may be found that the Seeker compulsively applies this "implant" as a filter to carrying out the directive to any PCL that they receive in session.

Records of confronting confusion—as held in memory as programming—usually are imprinted with "fear"; fear is the lowest denominator attached to implants that will keep an individual in the position of "effect." Fear is a "holder" on the Mind; it freezes the Mind in place with an inability to think analytically. This is so strong that many have had the experience of being able to *go outside themselves* during periods of intense fear/terror and watch just how unreasonable or irrational the response-mechanisms are during these periods. The *Self* is out of contact with the body because it no longer registers it as a safe POV to operate from.

When a problem is treated (emotionally) as a "Mystery" and then linked with "Fear" (being a "Mystery") it is easy to associate "fear" with the "unknown." By keeping personal *Awareness* in a low-level effect or POV, the true Self as Alpha Spirit is kept from achieving true realizations about existence. Where we have previously considered problems the overcoming of barriers or solving communication breaks, we may add to this concept the consideration that it is also overcoming a "mystery condition" or else an "unknown"—something that the Mind-System is wired to pursue naturally in order to occupy its own faculties. It sends its energies out to uncover the unknown mystery, but the Self is inhibited from confronting it and again finds itself in the middle of a confusion.

The "compulsive need and insatiable craving to *know* things" falls into this category; and so, things will be invented to be *known* about. The Mind-System is implanted to pursue data ceaselessly, which when operating compulsively, becomes an obsessive need or insatiable craving to simply "know"; and it doesn't matter what it *knows*, it just *has to know* "something" and there must be something more to be known because no solution is satisfactory to reducing the "unknown." Granted, this is a high level Alpha concern, but it is present throughout human activity in daily beta-existence at whatever point an individual is implanted to believe that "*To Know equals X.*" For many in the animal kingdom, "To Know—" might equal "To Eat" and so the better places to find better food might be the perpetual mystery to solve.

Of course, as we know now, the implanted impulse "To Eat" is not itself a *real* problem; but it *is* a problem implant, which for the purposes of the Physical Universe, produces an encoded effect on the individual that can be manipulated or imprinted on. This is to say that the "penalties" and "consequences" of the *game* of life in *beta-existence* is the "product of" the problem implants. These are merely activated[250] when the one-sided problem is confronted, such as "how to acquire food" &tc.

The manner in which an individual can confront, take responsibility for and willingness to be *Self-directed* will all determine their successful management of "problems." The simple event that a car is on fire is not a problem; the one-sided problem becomes "how to put it out" and other factors (sides) only add to this confusion when the situation is not managed properly and are allowed to be oppositional to simply accomplishing a goal.

---

250 **activating event** : an incident or occurrence that automatically stimulates a conscious or unrecognized reminder or 'ping' from an earlier *imprinting incident* recorded on one's own personal timeline as an emotionally charged and encoded memory; an incident or instance when thought systems are activated to determine the consequence or significance of an activity, motion or event—often demonstrated as *Activating Event → Belief Systems → Consideration.*

Our goal in *Mardukite Zuism & Systemology*—and the ministers and Pilots representing our paradigm—is not to simply "solve" all of a Seeker's problems for them. They don't want this anyway; the Alpha Spirit is a creative entity that likes to solve problems for *"purpose."*

Now, it *is* true that during professional Systemology Piloting, greater control of the Mind-System is temporarily managed by the Pilot for achieving greatest results from processing, but this control is then returned to the *Seeker* over the course of the session as they understand how to command it properly. Systemologists learn to solve their own problems better—managing and handling personal control of their Mind-System (and the Mind-Body connection) better—and in the end, simply *play a better game.*

:: 6 ::

THE GATES OF BETA-EXISTENCE AND BEYOND: PROBLEMS TRANSFORMED INTO POSSIBILITIES

It would be an acute simplification to simply push off onto the Seeker the idea that "they" create all their own problems—or at least the perception that things can be problems or that *Self* can be so unwilling to confront the nature of problems as to wind up in the middle of *confusion*. Whether or not this true, the actualized realization of this—and the ability to confront it with total responsibility as *Self*—is a very advanced "metahuman" state, which we would consider among the greatest apexes of the *Pathway to Self-Honesty*, or perhaps even superseding this. Such a *realization* cannot be firmly impressed with enforced knowledge; it simply lies in wait for when *Awareness* is high enough that the *Self* can handle it directly.

So then, what are we doing during the interim of this journey with all of these steps and procedures presented in *Grade-IV*?

First and foremost, we are introducing systematic methods that bring an individual closer to full responsibility of their own Mind-System—since they have been far too long under an illusion that everything that they are seeing in their reality and its mental imagery and all that associated encoding and supposedly logical reasoning is all being dictated, or otherwise externally determined; and this is one of the greatest tricks that has ever been pulled on the Alpha Spirit to make certain their consideration of *beingness* is entrapped within the confines of *this* beta-existence Physical Universe.

Inability to confront, approach and face the nature of present reality and existence, taking place in the Physical Universe and within one's own personal universe, is what triggers the automatic-responses and reactive mechanisms to be employed. The individual doesn't have the *Awareness* to be responsible for the present-time POV, so this is overshadowed with a lot of "suppositions" impressed as the "supposed to's" and "must haves" and other associative knowledge based on past circumstances. That which we have *lived* through is suddenly "safe" as a validity to our continued existence and survival—and this causes a tendency to form patterned reasoning; which if left simply to the old dead images[251] held in the Mind, you can be rest assured that these dead images can be infused with life and treated as a present-time reality—and the Mind-System will find no shortage of ways of justifying all of these reactions and associates based on its stores of experience.

As first began, back down the spiritual timeline of spiritual existence, the *Self* does not prefer to give attention (or "look") at things which it is not prepared to confront or be responsible for. There are old sayings regarding knowledge and learning that speak of the responsibility for what you "know"—and there is undoubtedly some truth to this, since we have discovered

---

251 **dead-memories** : outdated, inadequate or erroneous data.

—and relayed in the foundation of *Grade-III*—that "responsibility is power." But just as this responsibility deteriorated on a gradient scale, coinciding with the fragmentation of *Self* and the condensation of (and then entrapment in) *this* Physical Universe, so too can we systematically restore it by operating in a reverse order. It is *this* concept that has been ingrained behind mystical, spiritual and magical symbolism of various "*Gates,*" "*Dimensions*" and "*Levels*" in former traditions and cosmologies, but very often the truth of things is much simpler than graphically presented in Mystery Traditions; because what is a "dimension" really, but just a "POV."

> The next step at this juncture of the *Pathway* is to basically "flatten the wave" of "collapsed" considerations regarding "handling of problems," the ability to manage the Mind-System and adjust appropriately to change for optimum survival.

It would almost seem that these skills should have been entrusted to anyone thrust into the POV of a genetic vehicle in this existence, but what we have discovered is that these skills, techniques and various modes of instruction that we now employ for our Systemology and spiritual presentation of Mardukite Zuism are, in essence, "game changers"—and any time when such efforts have been demonstrated to any success in the past, it has usually been overrun by material-system corruption and the usurpation of less actualized individuals that use the same tech that could otherwise elevate the spiritual evolution of our planet but, because this *is* a "penalty-prison" fabrication of reality, these efforts resemble something more akin to a "jailbreak" than to a wide-spread rescue mission. Nonetheless, *Grade-IV* methods (as treated in the remainder of this volume) build upon the aforementioned instruction found in the *Grade-III* "core" of Mardukite Systemology[*] and also the *Grade-IV Professional Piloting Course.*[‡]

Δ Δ Δ Δ Δ Δ Δ

In earlier methods of our systematic processing, we present concise terms to our PCL, such as "Recall" in *Route-2 (AR-SP)* or even "Imagine" and "Create" as introduced in *A.T.* work; but in the *Grade-IV* methods, we have become accustomed to treating the energy ties to all of these methods as "communication channels" and the actual PCL (command scripting) varies based on the application. For our present purposes, and for introducing *Grade-IV* as *Wizard Level-0*, we intend to impress the conscious difference between *knowingly* creating (and commanding) the activity of the Mind-System and when it is based on reactive-response mechanisms (from the RCC) or other erroneous programming stored in the associative reasoning and logic applied within the Mind-System, which we must assume is not yet "perfected" if it is still in a state of fragmentation.

---

[*] Available within one single Master Edition hardcover volume as "*The Systemology Handbook*" by Joshua Free.

[‡] Available in the two volume set released as "*Communication and Control of Energy & Power*" (*Liber-2C*) and "*Command of the Mind-Body Connection*" (*Liber-2D*).

Therefore the methodology of systematic processing as we approach the *Gateways to Infinity* require a wider scope of PCLs. These involve freeing up the Seeker's considerations beyond "what has happened to them" already or their accessible memory recall of the same. These newer processes involve differentiating between what we are *knowingly* creating and willing versus what is set to go on automatic. Many of these skills—particularly those involving advanced applications of creative visualization and manipulation of mental imagery—are treated more intensely in the remainder of *Grade-IV* and beyond, but we intend a Seeker to progressively arrive at these levels as an evenly ascended stairway or ladder and not to place in view any goals that are outside the willingness and ability to reach.

In this first set of PCLs we are dealing with "concepts" and the intention and spiritual effort behind them on an Alpha level of interpreting the Mind-System data, which is to say "sense." We are not asking for any specific concrete imagery to come to mind, nor are we directly prompting a reactive response from former programming and imprinting (but should this come up—and it undoubtedly will—each point of turbulence should be processed appropriately before moving on); we are simply asking a Seeker to *get a sense of a concept*, and then we can see where this leads the Mind-System. In other esoteric terms, we are asking to "conjure" or "evoke" to Mind, because in actuality it is always *Self* that produces the *sense of a concept*—so we want to practice this knowingly.

Not all Seekers will have the same degree of fragmentation linked to the same imprinting events and interpreted in the same way as each and every other Seeker—but we know the basic outline of implanted programming and can apply systematic processing to all Seekers across the board. However, the individual responses to each stage of systematic processing is certain to be quite varied; and a *Pilot* is professionally trained to handle and treat the communication of these responses in very systematic way.[∞]

As a "Route-2" technique for *Grade-III* processing, the most basic session PCL would be: "Recall solving a problem" (or "Recall a time when you solved a problem") *alternated with* "Recall not solving a problem" (or "Recall an unresolved problem"). Because these were not found to yield very effective results for personal progress, they weren't included. These steps are graded for a reason. The practice of "imagining" *Self* solving and not solving problems worked marginally better, but without the type of training intended for *Grade-IV*, even this proved difficult in getting a Seeker to the point of optimum *realizations.* Therefore, the methodology of this present manual relies on an understanding of all former "Routes" of systematic processing, because elements of them all were included in order for a Seeker to "make the grade" so to speak.

These PCLs are applied following instructions given for *SOP-2C* and "Route-3"[*] and guidelines in *Crystal Clear* (*Liber-2B*). A Seeker/Pilot should work one "circuit" through (alternating the

---

∞ See "*Liber-2C*" and "*Liber-2D*"
* See "*Liber-2C*" and "*Liber-2D*"

positive and negative flows) until it can be handled with ease before working with the next one. Then after working through all the circuits, a return to the first one again is usually beneficial because it may be treated at a "higher level" of understanding than formerly. (This is generally true of most systematic processing because as a Seeker's *realizations* increase, so does their scope on reality.)

> Circuit-1:   Contact a sense of solving a problem.
>
> Contact a sense of not solving a problem.

Another example of this PCL could be stated:

> Circuit-1:   Get the concept of solving a problem.
>
> Get the concept of not solving a problem.

The idea of "contacting a feeling" or "conjuring the idea" is also acceptable for beginners. Each individual is going to find a pattern of speech that simply "resonates" better for them and will be most effective for systematic processing in early stages. But whatever pattern of PCL is selected, the same style should be used throughout each circuit. Other circuits of this processing cycle are as follows—

> Circuit-2:   Get a sense of (*terminal*)[∞] solving a problem.
>
> Get a sense of (*terminal*) not solving a problem.
>
> Circuit-3:   Get the sense of others solving a problem.
>
> Get the sense of others not solving a problem.

If a Seeker is working solitary on the *Pathway* (using *Grade-III* "Self-Processing" instructions given in *Liber-2B*[‡], they should work with any "surfacing thoughts" or realizations that come to mind. *Pilots* should be particularly interested in maintaining a flow of communication during processing to assist in a Seeker working out the necessary realizations for themselves before moving on to another process. It may be, without supplemental book instruction, that a Seeker in session will need to work up to the idea of even treating the subject of problems, for which they can increase their considerations by running through each of the following in series.

> Circuit-1:   What (is a) problem could you confront?

[∞] "Terminals" can be *any* communication terminal or any conception that can carry an emotional imprint or mental programming. Terminals could simply be the term "others" or "another person" or it could be identified more closely to a specific terminal (if it is a problem area), such a certain individual or the "phase" they represent, such as "mothers" or "teachers"—any specific "*is*" that the Seeker carries energetic turbulence with. See also *Liber-2C*.

[‡] Available as "*Crystal Clear*" by Joshua Free, and in the complete *Grade-III* Master Edition hardcover, "*Systemology Handbook.*"

Or using another style of PCL—

> Circuit-1:   What problem would be acceptable for you to confront?

Following the formula of *Route-3*, this PCL series continues—

> Circuit-2:   What problem could (another/terminal) confront?

> Circuit-3:   What problem could others confront?

It will be noted that even small "problems" are quite challenging for some Seekers to approach, simply because they are still hung up on basic semantics associated and encoding to even the very word "problem." For some Seekers, being directed with a PCL that implies confronting the entire problem is too steep of a gradient. This is one of the reasons we prefer to arrange *Professional Piloting* for processing through these Wizard Grades, because a solitary Seeker can easily become hung up on a part of the processing with no other expert advice or support on how to move through it. So as an alternative, the Seeker can consider (or be directed by a *Pilot* to consider) some type of problem and then inquire as to "what part of the problem" a Seeker could confront. It may be that even small problems must be confronted in pieces until an individual develops the certainty to manage the whole packages.

Δ Δ Δ Δ Δ Δ Δ

Individuals approach the field of problems based on a level of *Actualized Awareness* that they maintain. Any facet or directive could be developed into a problem, but that does not mean that they are in themselves the source of problems, such considerations again return to *Self* and their ability to manage Self-determined decision-making in *Self-Honesty* and free of the programming. At the lower spheres of reach within beta-existence, we have already taken great lengths to codify the nature of human problems and subject of their implantation. This knowledge or learning alone—without supplemental systematic processing—does not necessarily accomplish the necessary "release from condition" that we are ultimately after, but it at least leads a Seeker in the right direction for processing.

The first sphere of existence represents the lowest operable domain of *beta-existence* as this Physical Universe, which implants the problems of survival. The individual is a survivor and little more. In the second sphere, which is also within the "Reactive-Response" range of a genetic vehicle, we see even more built up on the subject of accumulation and loss, competition and scarcity. We find the development of the artificial personality and the ability to be encoded by deeper imprints of fear or threat to survival. All of this keeps an individual tightly wound in a small package, reacting against all conditions that are encoded to fear, abandonment, guilt and victimization. The being is now fully aware or conditioned to be aware that they are pitted against the problems of a material universe in order to secure the longevity of their own material survival as the end sum of their true spiritual POV—and herein we have discovered the lowest-level trappings of material existence; to get the Mind-System operating on reactive-response mechanisms which can otherwise be occupied to direct an individual between very rigid parameters of potentiality.

Since the true nature of the Alpha Spirit is a near-infinite Source-point of pure creative energy and activity, what passes for classification of a Mind-System—any part of the Mind-System, including MCC—must therefore be made up of some type of systematic "pattern" in order to operate as programming.

This means that even outside the domain of the RCC, we are still dealing with some kind of reactive-response mechanisms for an "analytical patterns" to form any kind of association in the upper-levels of the Mind-System. They may not be as crude as the reactive-response and fight-flight mechanisms of the RCC, but they are still patterned after some type of associated reasoning and thus full under the domain of a "mental universe" and not necessarily the true *Alpha* state of *Self.* As much as we would like to view the MCC as a kind of "higher Self" for the individual occupying a lower-level POV, it is certainly not the ultimate POV for *Self.* It is more commonly viewed as a threshold[252] between interior *beta-existence* and the *exterior* spiritual or "Alpha" universes.

Following a similar formula as former levels of systematic processing, a Seeker can conduct "objective processing" techniques to supplement subjective PCLs. In this instance, we want to practice with an arbitrary object—such as a *bell, book* or *candle*—that a Seeker does not have significant emotional encoding with (meaning an preexisting problem already). Eventually the same formula could be applied to processing out other actual problems, but the point of this exercise is simply to increase the range of considerations regarding the conception of "problems" and "solutions." After the object has been properly *identified*[*] (as a terminal) for processing, the following PCL are run similar to the first process in this chapter-lesson.

    Circuit-1:   How could (terminal) be a problem to you?
                    How could (terminal) be a solution to you?

    Circuit-2:   How could (terminal) be a problem to (other terminal)?
                    How could (terminal) be a solution to (other terminal)?

    Circuit-3:   How could others be a problem to (terminal)?
                    How could others be a solution to (terminal)?

---

252 **threshold** : a doorway, gate or entrance point; the degree to which something is to produce an effect within a certain state or condition; the point in which a condition changes from one to the next.

*   See *Liber-2B, Liber-2C* and *Liber-2D* for more information.

> Circuit-AT:[∞]   How could you be a problem to (terminal)?
>
> How could you be a solution to (terminal)?

Some of the answers to these may seem hard to come up with at first, but as considerations expand, even more ridiculous answers can be accepted as an increased realization that it is *Self* determining these evaluations, now and always. Another form of "objective processing" that may be employed between "subjective processes" could include the same or similar item with the intent that a Seeker invents ways in which the object is the answer or solution. In fact, as the former, this process can be applied to problem terminals just as usefully as it can be applied to creative practice.

> Circuit-1:   What problem could you have with someone with which (terminal) is the solution?
>
> Circuit-2:   What problem could someone have with you with which (terminal) is the solution?
>
> Circuit-3:   What problem could others have with someone with which (terminal) is the solution?
>
> Circuit-AT: What problem could you have with yourself with which (terminal) is the solution?

Δ Δ Δ Δ Δ Δ Δ

Naturally our ultimate intention is to get an individual to increase their willingness and ability to solve problems by understanding that they can just as easily create the conditions for the problem as the solution simply by reassigning their evaluations for things. This is a tremendous ability; true to the state of the Alpha Spirit—which in terms of beta-existence, has really only been understood by semantics of material social sciences regarding "intelligence." But obviously, there is more going on within the Mind-Body connection and the Mind-Systems itself than can be gauged on the most rigid standards of material thinking.

At the upper-levels of basic systematic processing—considered the "Wizard Grades"—the ability of the *Self* is very noticeably tied to "responsibility" and so we tend to emphasize that much more in the journey to regain and resume the original conditions for the Alpha Spirit to operate from as a true POV independent of the Human Condition. Of course, this does not happen all at once, but it is found to occur as a result of increased responsibility for energetic management of all the creative faculties attached to being an Alpha Spirit, in addition to all the other implants that have also been attached along the way. Although it is a steep gradient for a *Pilot* to thrust on a Seeker with their first pass through the processing prior to arriving at an *Actualized Technician* or "Wizard" status (which is a state that is "actualized"

∞ "Circuit-0" or "Circuit-A.T." is optional for the first pass through this present material (particularly for those still working through *Grade-III* instructions) and is included here for reference purposes only.

permanently by making a second pass through the processing while occupying this *A.T.* state, and thereby confirming the command of the Mind-Body connection for greater certainty of *Self* to go on to occupy higher POV without any ties that instill a sense of "loss" by elevating personal consciousness[253] to a higher universe of existence).

We are not limiting the application of this work to certain Seekers—hence its inclusion—but it is only recommended that PCLs for "Total Responsibility of Problems" be applied to those Seekers actually ready to confront the reality that they are the one's in the driver's seat for their existence. Once a problem (or turbulent channel/terminal) is contacted, the Seeker is prompted: "What part of that could you be responsible for?" And when that is no longer generating a response: "What part of that could you admit to causing?"

Since the nature of our "problem clusters" also seems to involve having two oppositional or separated POV, subjective processing can be supplemented with brief objective processes that include directing attention between two objects alternately, such as those methods connected to *"Bell, Book & Candle."*[‡] This is then applied to defragmenting previous unresolved problems by identifying the problem and then alternately spotting some point or facet from one side and then the other side—which is a further development on the suggestions given for *Route-1.*[*] This is particularly useful for reducing the energies holding together specific "problem cycles" that seem to reappear during the course of one's life.

---

253 **consciousness** : the energetic flow of *Awareness*; the Principle System of *Awareness* that is spiritual in nature, which demonstrates potential interaction with all degrees of the Physical Universe; the *Beingness* component of our existence in *Spirit*; the Principle System of *Awareness* as *Spirit* that directs action in the Mind-System.

‡   See *"Command of the Mind-Body Connection"* (*Liber-2D*).

*   See *"The Tablets of Destiny"* (*Liber-One*).

:: 7 ::
COMMAND OF THE MIND-SYSTEM
USING BETA-DEFRAGMENTATION

Prior to solidifying considerations and a POV for *this* Physical Universe, the "Mind-System" was constructed, created or formed—presumably by *us*—to engage with a "mental universe." The mental universe was constructed as a composite of personal universes, which have ever and always remained personal to the Alpha Spirit at its true state of beingness and existence, but from which the POV descended as more and more "communication barriers" were installed and more automatic mechanisms were established so that a "Mind-System" could actually exist and function.

> The "Mind-System" is no more the *Self* than the "genetic vehicle" or a physical body is. But, much like a physical body, it is used to direct information that the *Self* should otherwise have a clear observation of. When it does not, we say that this personal identity continuum is "fragmented."

Although we have "compartmented" our understanding of the "Mind-System" by dividing it into the RCC and MCC as a classification of function, the two parts actually do work together as a system as far s the Human Condition is concerned. So, up at least "7.0" on our Standard Model we have the Self as the Alpha Spirit in its original state, and down here at "1.0" we find an organic body that has roughly about the same solidity as the Physical Universe that it occupies as a body. We don't occupy that universe, but our considerations apparently do, and as we have moved down through a series of condensing universes, so has the parameters for the POV we agreed to and so with the vehicles we command. In between the functions and relative existences of these two states, we find a relatively large portion of the Standard Model—essentially all of, and concurrent with, what is classified a "beta-existence"—dedicated to the Mind-System. And it is *this* system that directly links the *Self* with *any* consideration of "form" as a beta-existence.

Somewhere along the spiritual timeline we received a distinct impression that we could be identified with a locatable form that could be harmed, that we could lose things, that we should protect our survival, &tc.—and while this may have originally taken place up in the *Alpha* regions of *exterior* beingness, it was *these* basic considerations on which the Mind-System was eventually formulated upon.

Circuitry of the Mind-System cemented through repeated validation of its use as a medium or catalyst[254] to evaluate experience and observations for us. And it is certainly no mystery as to how this could easily have compounded upon itself to set up a vast network array of

---

254 **catalyst** : something that causes action between two systems or aspects, but which itself is unaffected as a variable of this energy communication; a medium or intermediary channel.

channels and circuits to literally "compartment" its own systematic operations. This means that while the upper levels of the Mind-System may not operate at the same material reflex level as the "stimulus-response" and "flight-versus-flight" mechanisms of the RCC, it should be presumed that they at least must have *some degree of reactivity* to maintain any communicable preexisting programming or associative networking to operate functionally, and therein we discover the nature of what we call "implants," which underlie any all all other mechanisms, metaphorically being the very gel and membrane receptors[255] that make all other encoding, imprinting and programming possible.

As *Self* decided that more and more "things" were "unsafe" (not contributory to a continuing existence) to "look at" (given attention to), patterns of experiential memory developed; naturally leading to form "preconceptions" about things, which is to assign values or evaluate a reaction or response to a *past* "imprint" of something, which is then stored for future usage and treated then as "present-time" knowledge or experience. The only reason any being stores any information (or anything) anywhere is because of a potential for future usage. This is actually somewhat troublesome when we consider the amount of information stored from each circuit on a channel. These network chains of information will all be aligned to seem reasonable and yet, as suggested in the previous chapter-lesson, result in a sequence of cumulative "*must...*" and "*supposed to...*" implanted directives.

The issue in *beta-defragmentation* is the recognition that the entire "Mind-System" is still set up in a reactive way on some level. After an individual has repeatedly validated the lower range of POV as *Self*, the implanted problem network of directives in the Mindis not always clearly visible. It is, in essence, easier to demonstrate these principles using the electrical signals that are sent to muscles, or even in the range of emotion, where there is a great deal of solidity in sensation and its evaluation; but when it comes to the mental imagery and evaluations assigned to the same, it all looks so reasonable and logical from *down below*, as if any of the preconceived notions carry any data at all worth *knowing* about, and certainly not appropriate as a "filter-screen" by which we observe and interact with existence.

Elsewhere in esoteric lore, these "filter-screens" are treated as "*veils of existence,*" although fragmented knowledge throughout the ages left these traditions with no effective technology to effectively work through them, with all importance and significances generally assigned to corresponding "magical facets" and other "ritual specifics" with little regard for whether or not the mystic-magician or priest and priestess will actually arrive at the true *realizations* for *Self-Actualization* that are indicated for each of these "*Gates.*" This is what prompted our development of the Mardukite "Grades" and an advanced "Systemology" that could surpass all former "paradigms" and reach for ultimate goals of a *metahuman spiritual evolution*; thereby correcting the direction on the track that the standard-issue Human Condition has found itself (or in many cases, still not found itself).

---

255 **receptacle** : a device or mechanism designed to contain and store a specific type of aspect or thing; a container meant to receive something.

Since no preexisting system or paradigm of the past four millennium has demonstrated a motion in the direction toward correcting itself, we did not see it fitting to directly base our futurist evolution of Mardukite Zuism & Systemology on any one of them; with the exception, of course, being the consideration that we are an accelerated extension of the 20th century "New Thought" movement.

Systematic paradigms, traditions, religions and even material systems of the past have all kept attentions of the Alpha Spirit headed in the wrong direction—further entrapping *Self* into deeper and deeper material conditions of *this* Physical Universe—by promising (yet eluding evidence for) a greater evolutionary existence of the "spirit" by agreeing to more and more conditions of the Physical Universe and its systems.

While each of the paradigms may have been set up around a qualified source-point that *knew something*, the failure to adequately duplicate this original *knowing* across an organizational structure creates the mechanistic nature of the systems that ultimately leads to their breakdown. More attention is given to the repair and support of the failing system as an organization or paradigm than to the actual resolution of what is causing it; assignment of *cause* and the *responsibility* of creation has long since been lost to an unknown mystery—and so the existence of the "problem" perpetuates.

So now we have come to a point in the Mind-System where experience and familiarity lead to some quality of reactivity. An individual has "gone through" some experience or another and since they have obviously survived through it, there is some sense that there is something to *know* about it—and this of course creates a sense of familiarity with something, which we then can assign all kinds of associations and predisposed evaluations for. This could have started out very innocently and not as a means to entrap an individual, but ultimately that is what we see with its improper handling. It leads to associations that are limiting only for the fact that *Self* applies its own familiar experience from the past in order to experience the present—and where the "Mind-System" is concerned, this is done automatically by the very nature of its function when *Self* is not in "command," which is also to say "responsible" for handling a present experience. The "Mind" (or in many standard-issue cases, the "Body") is allowed to do the "looking" and "analyzing" because the Alpha Spirit no longer deems it "safe" (*willing, able, &tc.*) to do so directly.

Although we take up the subject of "POV" more thoroughly later on in *Grade-IV*, the important part to realize here is that all we are dealing with—whenever we are concerned with "layers" and "levels" and "*Gates*" and positions on the "ZU-line"—are various potential POV for *Self* (both *interior* and *exterior* to the Mind-System of the Human Condition), each with their own predisposed set package of parameters; and each, its own "game."

Δ Δ Δ Δ Δ Δ Δ

During the *Grade-IV Professional Piloting Course*,[*] we expressed the requirement for a *Seeker* to actually be "present" *in* a "session" in order for any systematic processing to be effective. That means that the *Seeker*, which is the *Self* of the individual, the "I" or "I-AM" as Alpha Spirit, must be "present" in the space-time *of the* "*present*" and able to experience it.

How is this even possible for anyone to achieve when there are so many "filter-screen-veils" of imprinted reactions and logical reasoning that stand in between the *Self* and the "present-time" reality to be experienced? It is *this* factor that led many philosophers into the field of "reason" and "knowledge" and the entire academic pursuit of "epistemology"[256]—because the nature of using a "sensation-based body" as the observer or measuring device of reality is always found to have serious flaws. Rene Descartes wrote about how the senses could be demonstrate to the fool the *Self* and therefore could not be trusted; yet a conventional practice or spiritual technology to remedy this effectively was still in need. And given the systematic religious and intellectual authorities[257] of this realm that we have always been up against, very few have put in the time—or believed they had the "luxury" to, during their lifetime in this existence—in order to work this all out...systematically.

If we could know for certain that the Mind-System was fully operational in *Self-Honesty*, then it might not be so bad to have some automated mechanisms doing some of the work for *Self*, but the information that it passes is always fragmented; mainly because it does not actually do any of the looking and experiencing and is instead dictating information back to *Self* based on the imprinted programming that has already been installed through experience. The *Self* is no longer doing any of the experiencing and is simply receiving a series of images on a screen that have very little to do with the present space-time taking place.

The entire "artificial" quality that *beta-existence* takes on is solely a result of this phenomenon. Of course this led some of the more "gnostic"-minded philosophers and traditions into developing a paradigm on the fundamental that the Physical Universe is somehow not real at all; which unfortunately removes *all* sense of responsibility for it from the individual. When we say it isn't there—setting up "communication barriers" with a universe along the way—we are leaving it up to something else for control of the communication of experience and reality for us.

One of the primary functions of the Mind-System that we know about quite readily from *Grade-III*, is the evaluation of personal knowledge to estimate the application of "effort"—which is sometimes interpreted as "force" in the Physical Universe.

---

[*] "*Communication and Control of Energy & Power*" (*Liber-2C*) and "*Command of the Mind-Body Connection*" (*Liber-2D*).

256 **epistemology** : a school of philosophy focused on the truth of knowledge *and* knowledge of truth; theories regarding validity and truth inherent in any structure of knowledge and reason.

257 **authoritarian** : knowledge as truth, boundaries and freedoms dictated to an individual by a perceived, regulated or enforced "authority."

The only true directive force that we know of in this universe *is* the Alpha Spirit, but its own essence is not *of* and *as* this beta-existence and thus is not able to be defined by any of its parameters of weights and wavelengths. Therefore, the *Self* has set up channels of communication between relays and control centers that make this estimations of effort based on past experience. This "analytical pattern" in itself becomes something of a reactive response mechanism that can operate as a "push-button" system—and that is very detrimental to maintaining *Self-Honest* clarity of the Alpha Spirit as fully *Self-determined.*

> An interesting correlation between *beta-Awareness* and the estimation of effort to create an effect is that we have found that as one moves down on the *ZU-line* (in *beta-Awareness*) they feel the need to exert more and more effort at the POV level they are occupying in order to earn the effect at that level.

You can demonstrate a registry of this in the "emotional state" an individual is suspended in at those levels where it is conceived that more and more effort will be required in order to create a change of state—even violently, if necessary—until an individual (upon failing to accomplish their ends at each point) finally succumbs into being the effect of oppositional forces, such as the basic material of the Physical Universe; *if* and only *if* it is perceived as maintaining a greater level of command then the POV that *Self* is fixated at.

Of course, when this registers fully as being "total effect of external forces," the Alpha Spirit withdraws completely from the Mind-Body connection; we generally speak of such "genetic vehicles" as being dead and abandoned. The physical aging cycle could even be considered a reduction of high-level *Awareness* "POV" (as cause) being transformed into an effect. The individual eventually decided they could do no more to create an effect and be responsible for no more cause in this existence; so the Alpha Spirit just stops "looking" at the organism goes more and more on autopilot (having accumulated more and more "experience").

A *Pilot* should note when—during the course of this *Grade-IV* methodology—that the Seeker can also come to a realization of a problem simply being a "no-problem" or even a solution that seeks or creates problems. This comes up more frequently when imagining, inventing or creating problems as an energetic practice. Anything willingly and knowingly created within the personal universe can be controlled as "ownership" or "having" (which is again, to say, "responsibility")—and there is no need to worry about having to hold on too tightly because we can throw these away and easily create or duplicate any former creation again. Scarcity of this *Awareness* is part of what causes *Self* to "hold on" so tightly to traps in any material *beta-existence.* As a Seeker comes to an increased realization that a problem can be easily created at will, then there is less of a drive to remain suspended between existing ones.

The "Mind-System" is designed to essentially copy images of the universe in which the *Self* is extending its POV, which for our purposes now, has been restricted to the "games" of the Physical Universe (*beta-existence*). Since there is any number of creators putting forth their energies in this objective reality, it can be assumed that from the perspective of the "genetic vehicle" and Mind-System that the *Self* is not directly responsible for the "creation" of the other-determined existences encountered in the Physical Universe. It does, however, begin to make copies of images that it encounters, which along with encoded programming and evaluations, is stored in the Mind-System as a representation of the external world. It is the contents of *this* Mind-System that an individual *is absolutely* responsible for—and it is *this* contents that contributes to experience of a personal universe that esoteric philosophers in the past have referred to as the "mental plane." We could go on to include the "astral plane" within this domain, because our mystical experiments revealed that even traditional "astral work" does not liberate *Self* from the "stuff" contained in the Mind-System. A "magician" is still very much operating from a POV within the *interior* of the Mind.

In most cases, *Grade-I* pursuits on the "Route of Magick & Mysticism" are not the most effective in achieving the highest *realizations* toward the ideal state of being for *Self*; even if their original underlying purpose was to get there, most simply don't. In most cases, the remedy for material problems (at low levels of operation) is a solution that in itself creates more problems. The root of problems and the Mind-System is never fully uncovered, only demonstrated—and repeated lifetimes of use, with a practitioner suspended at the *First Gate* —simply validates and reinforces more and more of the mechanisms present in the Mind-System.

As the subject of "problems" becomes more interesting to lower-level understanding than pursuing "*Self-Honesty*," an individual spends more and more of their directed attention on simply finding more creative ways to remain suspended in lower level considerations and driven by the same implanted problems as the uninitiated.

One reason our advanced work is built upon the *Grade-II* "Route of Mardukite Mesopotamia" is because the only effective elements from *Grade-I* that lent any assistance to achieving access to the true *Second Gate* were all connected to the highest pursuits of mysticism in the direction of "Divine" and "Celestial" branches of magic and mysticism, including Druidism—which may have been repeated in their own fragmentary designs throughout history, but of which, in their purest form, can be traced back to the systematization of the ancient world in Babylon, by the Mardukite Babylonians.

Although many who have delved into esoteric and mystical pursuits are generally familiar with the more recent Judeo-Semitic interpretation of the "Kabbalah"—upon which the last two millennium of "ceremonial magic" is strongly based—even a casual amount of research into the much more antiquated *Grade-II* presentation of "Stargates" in Babylon (or "*Babili System*") easily demonstrates that the more commonly known "Kabbalah" is an importation of older "stronger" Babylonian lore.[*]

---

[*] See especially the *Grade-II* volume, "*Practical Babylonian Magic*" by Joshua Free (also available in hardcover as "*Necronomicon: The Anunnaki Grimoire*" and within the complete Grade-II Master Edition

If we were to follow the track laid down by the *Ancient Mystery School*, then:

The *First Gate*, being the lunar level, would correspond directly to pursuit of magic, mysticism and enchantment discovered in *Grade-I*, as an initiate moves up into the various "elemental" dimensions of the Earth Gate and beyond, tapping the first veil;

The *Second Gate*, is of course linked to Mercury and the hermetic spiritual quasi-religious styling of *Grade-II*; and in ancient Babylonian Mardukite tradition, this position was correlated to the Anunnaki demigod "Nabu," the systematizer of Babylon, developer of the stylus and refiner of cuneiform script in order to structure the first "priesthoods";

Which brought us to the forefront of the *Ishtar Gate* in *Grade-III*, the Venusian veil that most have never successfully crossed beyond from the Human Condition, because it requires attainment of *Self-Honesty* to the degree that reactive emotional ties to the lower realm are resolved;

As are the higher analytical systems of the Mind, which are confronted and disintegrated directly at the *Fourth level*; the Sun...[*]

Such information is provided solely as reference to the esoterically inclined that may have missed this basic structure of ascent up the *"Ladder of Lights"* that is present in our interpretation of the journey for Mardukite Systemology.

Δ Δ Δ Δ Δ Δ

We have, at this stage, uncovered most low-level *beta-mechanisms* of the Mind-System that seem to "key-in" a "push-button" response association with reality, and its communication, which contributes greatly to an entrapment to the Physical Universe by consideration. There are many ways of working to undo these mechanisms—most of which continue to be explored throughout the forthcoming manuals covering these upper-level "Wizard Grades" of Mardukite Systemology. However, there is *more* instruction and practice that is critical here at *this* stage, before we leave an ambitious Seeker (or *Pilot*) off on their own to go treating "all the problems of the world."

We've come to realize that "problems" are simply a part of the "game" scenario; without them, you don't really have any conditions for which to operate a "game"—there doesn't seem to be any real purpose without something to "solve." This is confined to parameters of whatever "game" is in play—and the most unfortunate event for the Human Condition is when it stopped *knowing* that it was playing this "game." Did you ever play "cops and robbers" as a child? You would invent a set of rules and parameters; might even take turns—

---

hardcover *"Necronomicon: The Complete Anunnaki Legacy"* edited by Joshua Free.

[*] Which should prepare one adequately for the *Wall of Fire* one must pass through to get beyond the *Fifth Gate* circumvented by "martian" and "martial" energy.

playing one side and then the other. It was a "game," you knew you were playing it, and so did your friends. But maybe after a while, maybe that wasn't interesting or challenging enough to occupy attention, or be considered "fun."

> So, what would happen if, in the middle of this game, you forgot it was a game and all you knew was the identity role persona that you had taken up to play it? What then? Well, this is a fair approximation of the current state of the Human Condition; and what's more—the *Self* is left to remember and rediscover all that it chose to forget to play *this* game. It's high time we found a final resolution to *this* one, so we can play a better one. If you haven't noticed yet, there is no way to "win" *in* the Physical Universe, but we can uncover the rules and break the chains to when we first agreed to occupy a Will *under* Law.[258]

We have discussed the nature of lower universes, such as the one we occupy, as penalty-prison systems developed beneath a higher order of POV and existence—and this sequential deterioration has continued to go on for some time. Many have an innate sense of the "Otherworld" or the magical lands of "Faerie" and other elements of medieval-themed fantasy, which seem as if they are a memory of a former time on *this* planet, but they are not. In fact, even when such periods of history *were* taking place prior to the industrial mechanization of society, they were themselves developed on and meant to stimulate memories of this *other* universe. Such made *this* version of Earth more resonant and familiar with the immediately preceding "home" from which we came. After all, it is up to the inmates to create and maintain their own standards within the cells. So, naturally there is plenty to *do in beta-existence*, and countless ways to occupy attention during one's imprisonment—but to what ends?

For those that have risen above the lower-level of sensation—those that we say are wholly living reactive lives based on immediate gratification of pleasure centers in the body—there are just as many, if not more, "intellectual traps" installed into the Human Condition once the POV of *Self* moves down into the range of this Physical Universe. Most parts of the Mind-System are simply driven by the discovery and craving for "new information" in an attempt to understand whatever previous layers of information have already been collected. It is ceaseless in this pursuit and stores information across many lifetimes, so it is unlikely to be localized[259] in the "brain" although it appears to use that physical organ as a communication control center for the genetic vehicle as part of the RCC relay system of the Mind-Body connection.

Accumulation and storage of new information by the "analytical" database of the Mind-System is always appropriated an evaluation based on what has already been stored along appropriate "nodes" of that subject (as associative data held in the Mind). When this is left

---

258 **Cosmic Law** : the "Law" of Nature (or the Physical Universe); the "Law" governing cosmic ordering; often called "Natural Law" in sciences and philosophies that attempt to codify or systematize it.

259 **localized** : brought together and confined to a particular place.

completely on automatic, outside the control of the Alpha Spirit, the tendencies and patterns of the circuitry becomes fixed—thereafter the individual is unable to consciously adjust their POV or considerations from a strongly impressed set of operations that they stopped being responsible for. The true stability of a Seeker—at least as it relates to *Grade-IV* guidelines—is measured observably by their ability to handle the problems and solutions of life, which is undoubtedly reflected in the communications during a systematic processing session.

An individual's own point of stability is essentially the fundamental implants and premises that they will eventually group all of their other accumulated associated data around. Their freedom to create and un-create within the Mind-System is the focus of all systematic processing whereby the Seeker is directed to generate, imagine or create various concepts, ideas, scenarios, and flows of energy along any channel, all freely at will. As a related area of focus, we are also dealing with the concept of "change" because all problem solving is an innate property of movement—or any growth or continuation along a line or track. Even if the "problem" to be solved is simply only stated as getting from "point-A" to "point-B" there is still *some game* to play; something to solve and *do.* Of course, in the Physical Universe we have misaligned our priorities of beingness and repeatedly substitute *doing* anything in place of *knowing* something real (as a factor of "purpose" in movement).

:: 8 ::

COMMAND OF THE MIND-SYSTEM
FOR A METAHUMAN EVOLUTION

Most continuing Seekers, Mardukite Zuists and Systemologists are familiar with our use of mental exercises referred to as "systematic processing." Several examples of this are found in former lesson-chapters and an entire introduction to its use may be found in *Grade-III*.[*] A professional course in systematic processing and Professional Piloting opens the present *Grade-IV*.[‡] What we are approaching in *Grade-IV* is a heightened *beta-Awareness* that when compared to the standard-issue nature of the Human Condition, would seem quite *metahuman*, quite simply because the individual is not rigidly fixed on a POV within the biological or intellectual confines of this Physical Universe. The gradient of work that is before us is bordering on an even greater work that will be developed further in a forthcoming volume,[∞] which will place greater emphasis on the command of thought and mental imagery—but there are a few steps we must still climb before reaching that point.

The common denominator of all advanced pursuits by an individual, whether physical mechanics or esoteric mysticism, is the handling of "energy." Energy, as we have found, is the fundamental unit of whatever is taking place in existence—any existence. The practice of personal control and determinism over the control of energy has been the subject of pursuit by every scientist, magician, priest or priestess, separated only by their semantic applications to observable levels of energetic interaction. This matter of "energy"—and the subject of "ZU" as a semantic—is best explored in the introductory *Grade-III* lecture series titled: *"The Power of Zu."*

It would seem, as with many subjects within our Systemology, that the ability to handle the energy we face everyday and its ideal use to increase our own spiritual abilities (those that extend to include more than this lifetime or else beyond *beta-existence*) should be a commonplace education; and yet we find it nowhere expertly given in any universal applications, with the lowest possible understandings of it rendered as "science."

There is nothing inherently wrong with a "science"; it seeks simply to *Know*, but it also must invent its own knowledge to *Know* about since it has eliminated from its equations the only lifeform that can actually do any *Looking* and *Knowing*, which is, of course, the Alpha Spirit in its true and defragmented POV.

---

[*] The concept is introduced in *"The Tablets of Destiny"* (*Liber-One*) but is explored more extensively in *"Crystal Clear"* (*Liber-2B*).

[‡] Referring to *"Communication and Control of Energy & Power"* (*Liber-2C*) and *"Command of the Mind-Body Connection"* (*Liber-2D*).

[∞] *"Liber-3D"* (*forthcoming*).

Others have made attempts at getting to a point of "more than human" as well, but nearly all the current efforts are directed toward the external transhuman[260] technologies that really do not serve a purpose in achieving the actual state of beingness we refer to as "*Homo Novus*"—and therefore, to differentiate from these materialists, have attributed the main brand of our *Grade-IV (Wizard Level-0)* work as "Metahuman Systemology." This state appears to be primarily attained by *beta-defragmentation* of the "Mind-System."

We can easily observe that the "Mind-System" operates on implants, directives and impulses —which some modern philosophers have even equated to electrical activity—all of which contribute to personal fragmentation when allowed to operate *unknowingly* based accumu- lated imprinting as a communicator of reality to *Self.* It is obvious that this system is not stored within and as the genetic organism, quite simply because there is not nearly enough cellular space—even in all the regions of the brain—to actually account for the amount of in- formation an individual stores on their personal identity continuum (which is experienced in relation to what we refer to as the *ZU-line*).

The accumulation of energetic "mass" in the form of imprinting and programming through the course of many lifetimes has led a once god-like spiritual being into the position of carrying a lot of "baggage"—all of which has increasingly weighed the *Awareness* of the individual down into the agreements of *this* Physical Universe due to an inability to unload these burdens along the way. It seems as though we felt we really needed to carry this weight with us in order to "have" something; not realizing that we could have released ourselves from the hold and just created it all again later.

Mental images themselves do not seem to weigh us down—because they may be easily dispersed and created if Self-determined; only associations that are tied to them cause us the fragmentation—whether from emotional encoding or mental programming. It seems that at this point of the journey there are at least enough individuals now starting to take notice of this to make it a "thing" and the subject of a "New Thought" movement that is, in the words of this generation, working to liberate Alpha Spirits from the *Matrix*; freeing considerations to a higher POV that it has otherwise agreed to be shielded and blocked from in order to experience *this* "down here."

So, what have we learned about the interaction of the Mind-System in relation to the Physical Universe that is useful on this journey of *beta-defragmentation?* We have learned that the electrical charges directed from *Self* as "ZU" (*Spiritual Life Awareness Energy*) operate quite

---

260 **transhumanism** : concerning the next evolved state of the "Human Condition," which is to say either in a direction of "internal" or "spiritual" technologies that advance the *Self*, or the direction of "external" and "physical" technologies that either modify or eliminate the *Body*. In our present state of society, it is the "physical" that is selectively *sold* to the masses so that only a select few may experience the "former"; in *NexGen Systemology*, also referred to as "metahumanism" with an emphasis on "spiritual technologies" as opposed to "external" ones.

differently in *beta-existence* than they do in the Alpha zones of a spiritual one. In fact, there are many instances where it seems that the Physical Universe is designed, being the expertly prison that it is, to deliver the opposite results to what we are after, all of which is related to the fragmented problem-implants that we have illustrated and described in this present manual.

For example: have you ever noticed in your experience with the Physical Universe that when you really really want and crave something—practically to the point of obsession—it always seems just a little bit out of reach? How about avoidance?—Did you ever notice that when you put up resistance to avoid something too strongly that you end up with it staring you right in the face? It's rather like getting a bit of dust or particle matter out of your cup of coffee: you reach at it with your finger and the wave-force distances it away from you; then when you withdraw it returns to where you had originally targeted.

Given how low *Awareness* levels tend to reach in *beta-existence*, it is found that an individual believes they need to apply an excessive amount of effort and force to create the change they want. This is, of course, after experiencing many diminishing universes that each became just a little bit more "solid" than the one before. Even the material universe of the "Other-world" or "Magic Kingdom" that immediately *precedes* or *envelopes* this one (depending on your opinion of dimension) is not that much different in terms of solid forms, except that consideration for the power of the electron is not restricted to a "wire" which is therefore what makes "magic" possible there. And most of us have had some sense or memory of such a place, but it will not be found on this planet, only those times when those sharing this same memory have again made attempts to create a facsimile of it here.

△ △ △ △ △ △ △

Obviously we all have the ability to create, we have the ability to apply will as intention, and certainly not every effort or reach that we extend in this existence is thwarted;[261] but somehow the important ones—at least the important ones to defragment your personal wiring—are placed in the category of "implanted problems" which are not otherwise *Self-determined* if following standard-issue programming. So, the individual that craves something to the point of obsession is wired to never receive it so long as they continue to apply those patterned efforts. For all of the hard-wiring that is attached to the response-mechanisms of the genetic vehicle, energy does not respond very well to a source-point of erratic desperation.

If you consider the projection and reception of energy as the force and suction demonstrated on the speck of dust in coffee cup, it is easy to see how we again return to the intention-counter-intention cluster of energy that we have formerly referred to as a *real* "problem." In the case of an implanted problem, these channels are encoded to exhibit polarities of "must have" and "can't have" on a particular terminal, and unlike a traditional "problem" cluster,

---

261 **thwarted** : to successfully oppose or prevent a purpose from actualizing.

these implants can actually create a vacuum of energy by the directive that we "must" draw in and keep away various currents in order to satisfy programming.

Most materially successful individuals that have achieved success via their own actualization have realized that the key to it all is "acceptance"; which someone maintaining a much more fragmented Mind-System would only be able to realize as "indifference"—and they are not the same. In fact, we have a place for "indifference" on our *beta-Awareness* scale, and it is not very high. Acceptance, however is a much higher state that revolves around a freed consideration of potentiality that they can "have things" without being obsessive or compulsive about the efforts to attain them. This is, of course, also relies heavily on one's own *determination,* but that again is not the same as the repetitive application of an effort against opposition; it is instead a directing property of *Will* and *Intention.*

Exercises where an individual is maintaining their own control of their mental imagery may be used to assist the fluid willingness to "have" and "not have" without emotional reactivity. Even a casual disinterest is better than supplying more energy to rejections, blocks and barriers. As many have already found, where it comes to emotionally dramatic displays of behavior, it is simply best not to engage. Although the direct handling of creative imagery handling in the Mind-System is relayed in greater detail later on in *Grade-IV,* for our present purposes, we have been using the PCL technique of "getting the sense of" something, which is especially useful for those Seeker's who do not yet already have an existing demonstrable management of their own mental imagery in which to employ for systematic processing. This too is developed with further practice along the way.

In systematic terms were are speaking here in terms of "acceptance" and "rejection"—and we can find this semantic present in our processing when a *Pilot* asks what a Seeker could "accept" about something, such as the situation they are having a problem with, or even something in the room (as practice), and then what could they "reject" about it, alternating with each PCL. For example, in an objective process for *"presence in space-time,"* we could prompt: "What about this room could you find acceptable" alternating with "What about this room could you reject." The point here is not to necessarily *have to* accept or reject anything about anything, but that the practiced *Self-determined* consideration for this fluidity remains unfixed.

We can introduce some practice in applied visualization for those that *do* operate easily at that level already, but these same considerations of energy management *may* be contacted with "getting the sense" or "concept" of something—and, of course, if any automatic reactive-responses or mental images *are* triggered by working with a particular circuit, channel or terminal, this should all be noted within the session journal or Pilot's log.* Keep in mind that

---

* There are specially prepared *Systemology Adventure Journals* or *Flight-Logs* now available specifically for recording the various facets, conditions and tech applied during the course of systematic processing sessions.

anything that *is* uncovered, should be handled on the spot since it is obviously a "problem" area that the Seeker is prepared to face.

Any "thing" or "concept" that a Seeker is compelled toward and uncovers as a source-implant of directive toward something they aren't achieving can simply reverse the charge at will, by "wasting" or "throwing away" mental images and other imprinted symbols that key this cycle in. It is apparent that the Mind-System has treated it as a scarcity and therefore has encouraged hanging on to it just a little too tightly. This can also be practiced further, systematic in a processing session by "imagining" or "getting the sense of" the free circulation of these energies, which would run along the lines of:

Circuit-1:   Imagine (get the sense of) giving X away to another.

Circuit-2:   Imagine (get the sense of) another giving X away to you.

Circuit-3:   Imagine (get the sense of) another giving X away to others.

Another alternating PCL technique to assist in freeing up the considerations surrounding *"having to X"* as the only solution or answer is—"What could X be a substitute for?" and "What could be a substitute for X?" Also returning to our previous alternation[262] of consideration—"What X could you accept?" and "What X could you reject?" Or, if more applicable for lower levels: what *about* X, &tc. You can practice this in regard to aesthetics or survival with any terminals that can be alternated with "desirable" (wanted) and "undesirable" (unwanted).

In addition to what we have presented, the "must have"/"can't have" problem implants are equally found in the "must avoid"/"must keep" programming. In the same way that the "having" is most strongly linked to circuitry pertaining to *loss*, the "must avoid"/"must keep" implants are often related to registries of *pain*. This is logical, because "pain imprinting" is most solid of the reactive-response encoding, so its memory is intended to remind an individual of what they "must avoid" by giving them heavy pain imprinting they "can't get rid of." This is similar to the previous issue with the exception of the direction of energetic flow held in suspension.

When we are stuck in the "must have"/"can't have" cycle, the subject or concept (terminal) is treated as a scarcity and so the energetic solution is to imagine and create an excess of positive mental imagery that can easily be discarded, thrown away, wasted or dissolved without an emotional response engaging to "hold on" to it. After this is resolved, a Seeker is instructed to work in both directions of energetic flow, practicing fluid "acceptance" and "rejection" by alternately "throwing away" the images *and* "pushing them in" on the person-

---

262 **oscillation-alternation** : a particular type of (or fluctuation) between two relative states, conditions or degrees; a wave-action between two degrees, such as is described in the action of the *pendulum effect*; a flux or wave-like energy in motion, across space, calculable as time; in systematic processing, alternation is the shift between two direction flows on a circuit channel, such as *inflow* and *outflow*, or between two types of processing, such as *objective* and *subjective*; alternation of a POV creates "space."

al universe. When we are treating the problem implants of "must avoid"/ "can't get rid of," the procedure is run in reverse, getting an individual to accept more of it by "pushing" an excess of the imagery into the personal universe until it can be "accepted" or "rejected" without fragmentation. This is not an instant process, but it may be accelerated with proper handling of mental imagery.

Δ Δ Δ Δ Δ Δ Δ

What we have done in thus far in the manual is circumvent various layers of the concept that we could simply state as "change." We are, throughout our Systemology, referring to various "fluid" conditions, freedom of willingness and the various POV whereby *Self* can get "stuck" in. All of these surround the idea of "change"—and the ability or willingness and the inability or inhibition to "accept" and "reject" *changes* without emotional turbulence or other mental fragmentation. It may be that one of the closest points the Western[263] world achieved in understanding this is referred to as the "Serenity Prayer," which is in more common use among a popular *Alcoholic &tc. Recovery Group* (we won't name directly) than anywhere else in society. It speaks strongly of the willingness to accept conditions and ability to change them; with the wisdom to *know* when to do each.

We have generally found that higher actualized *beta-Awareness* equals greater ability to *Self-direct* change. This also does not mean that things have to always be changing and that some things cannot remain as they are—but again we are concerned with the free-flow of energy in either direction. However, when it directly concerns the patterns of imprinting and programming, we see a greater tendency to resist change, because energy is holding the POV of *Self* in a suspension point that elusively provides an illusion of a similar near-stasis point of the actual Alpha Spirit ("I").

Very often as an individual runs further along in life on their reactive-response mechanisms and other mental programming, the command of these functions is replaced by automatic patterns. The information from painful and critically stressful events or other perceived threats to one's own environment or "person" is carefully "photographed" and stored as an imprint with all of the *facets* treated equally to the emotional encoding of the experience. And the individual will "hold on" to this imprint very tightly believing that by carrying the full intensity of this experience they will "have" something to "know" in order to keep situations from getting any worse, or at the very least, being allowed to "repeat/duplicate" again. Well, this would all be fine and good if it were not encoded by reactive-responses and then thrown back up at us in the same way, filtering out the true knowledge and experience and *presence* of *Self* in the *present* by bringing "dead images" back to life as the *present...* and what strange *Necromancy* this is.

---

263 **Western Civilization** : the modern history, culture, ideals, values and technology, particularly of Europe and North America as distinguished by growing urbanization and industrialization and born from a rebellion to strong religious indoctrination.

Ability to manage "change of energy"—which essentially qualifies to include *all motion* of energy—is characterized by a willingness and certainty to channel energies, which is *not* here used in the same sense that the mystic may have referred to channeling entities in the past. The "channel" here implies application to "communication" semantics, regarding personal energy and its interaction with the external world. This means the *Self-directed* intention to "hold energy" and "move energy"; which we can apply demonstrations of as "objective processing" in material existence. Here, the *Pilot* may apply the *"Bell, Book & Bottle"* objective processing kit to a session; having the Seeker select an object and reach for it to hold it still several times; then selecting an object, the Seeker would repeatedly practice holding it to make it more solid (by intention); and finally, selecting an object, choosing to move it to another *Self-determined* location and then doing so. This would be a basic objective systematic processing example regarding "change." The same formula could be practiced on parts of the body, such as an arm or leg, &tc. (including trouble areas—since an individual is "out of communication" with those), and also in mental imagery exercises.

"Change"—and *alteration* in general—is a basic property of systematic control. Other than the initial start and stop of a flow or circulation of energy, about the only other thing energy actually does within and as any system is "change." Even the motion of energy itself is a "change" in state or location, which creates a flow-pattern that we measure in cyclic waves and generally define as "time." The two sides of change, as we have seen, regard holding something still (and keeping it from going away) and then the motion of enacting[264] a change. We are, of course, looking for acceptance for both, free of reactive-responses and fragmentary energetic turbulence. The concept of change may also be introduced into processing as a "terminal" using any of the available "Routes" for *SOP-2C*.[*] For example, a Seeker could use "Analytical Recall" (*Route-2*) or "Contacting Communication Channels" (*Route-3*).

> BASIC CHANGE PROCESSING—ALTERNATING (GENERAL)
> Circuit-1:   Get the sense of changing something.
>   Get the sense of stopping something from changing.
>
> SYSTEMATIC PROCESSING OF CHANGE (AR-ROUTE-2)
> Circuit-1:   Recall a time you changed something.
>   Recall a time you stopped something from changing.
>
> SYSTEMATIC PROCESSING OF CHANGE (AR-ROUTE-2 EXTENDED)
> Circuit-2:   Recall a time when (*terminal*) changed something.
>   Recall a time when (*terminal*) stopped something form changing.

---

264 **enact** : to make happen; to bring into action; to make part of an act.

* See *"Communication and Control of Energy & Power"* (*Liber-2C*) and *"Command of the Mind-Body Connection"* (*Liber-2D*).

SYSTEMATIC PROCESSING OF WILLINGNESS TO CHANGE (ROUTE-3)

Circuit-1:	What would you be willing to have changed in "another"?[‡]
Circuit-2:	What would you be willing to have another change in you?
Circuit-3:	What would you be willing to have another change in others?
A.T. (*Opt.*):	What would you be willing to change in you?

The Mind-System has a unique way of collecting and registering its data; in both instances, acting as a mirror and crystalline lens between the "observer" and the "observed." These mechanisms all act to create, store and display "mental images" which are essentially *facsimiles* or *copies* of what is registered as experience of existence. The motion of energies captured on these images is what gives a sequential record of "time"—and which may be accessed as memory.

The Alpha Spirit, from it stasis spiritual position, engages its attention by projecting toward "mental images" of motion (as a terminal), which then produces energetic activity. The differential[265] between a static point and the image of activity is what creates the "charge." We are, in essence, imbuing a certain degree of "life" and "beingness" into these images, including all of our previous imprinted records of the Physical Universe. A fixed association of emotion or knowledge (both are source-points of *knowing* at their own levels) to any of these "charges" or "terminals" is what can get an individual stuck in patterns that they are either unwilling or unable (since the two are virtually the same) to "change."

An individual goes along thinking and acting as though such and such just *must be* the case so much, even against all odds, that they are unwilling to channel any other energy free-flow. They find themselves coming up hard against objective reality and other energies with repeated invalidation. Maybe the individual *is actually* correct in their knowledge, but what does that matter if the emotional and intellectual ties to this condition are destroying the individual from the inside due to their inability to release the hold on it. The mere unwillingness to *be* in any other POV is going to give them a difficult situation to manage and contribute to an existential demise; because we know that the willingness and ability to manage and adjust to all conditions freely is the key to our continued survival and upward progression.

Δ Δ Δ Δ Δ Δ Δ

Elsewhere in our Systemology,[∞] the mental imagery that is captured during times of pain, loss and other perceived threats to survival are referred to as *Imprints.* Unlike the associative

---

‡ This terminal is generally treated as "another person," though it could also be treated as any other "lifeform" or even the Spheres of Existence. If there is a problem terminal being processed, that can be used as well.

265 **differential** : the quantitative value difference between two forces, motions, pressures or degrees.

∞ See *Grade-III* materials, particularly "*The Tablets of Destiny*" (*Liber-One*) and "*Crystal Clear*" (*Liber-2B*) also collected in the complete *Grade-III* Master Edition hardcover "*Systemology Handbook*" by Joshua Free.

knowledge and stored memory used in the analytical range of the Mind, these *Imprints* have very little rationale to their content. Any and every *facet* contained in one of these heavy *Imprints* can be given the label "must avoid" including locations, individuals, sights, smells, tastes, sounds, humidity, lighting…and just about any other possible *facet* of perception received at the moment of *Imprint,* which is an *Imprinting Incident.*

This fragmentation (automatic response mechanism) is later validated when stimulated by the environment, called an *Activating Event,*[266] and then treated based on the content of the original *Imprint,* including any manner of sensation and emotion and pain that can be triggered by the RCC to alert the identity that they "must avoid." This develops into an automatic "flinching" or "recoil" any time *that* channel of communication is contacted. It should be understood that the internal mechanisms of the Mind-System are only stimulated into action by the environment; all of the actual energy contained, entwined and solidified in the emotional imprint as a "mass" is being supplied by *Self*—and since this is all happening automatically, the solution is to return the Seeker to a point where they can manage such products of the Mind-System clearly on their own determinism.

CHANGE/UNCHANGED SUBJECTIVE PROCESSING (ROUTE-3)

Circuit-1:	What *could* you change?
	What *would* you leave unchanged?*
Circuit-2:	What *could* change you?
	What *would* leave you unchanged?
Circuit-3:	What *could* change others?
	What *would* leave others unchanged?
A.T. (Opt.):	What *could* you change about you?
	What *would* you leave unchanged about you?

Objective processing may be applied using the same formula, simply having a Seeker look around the room, locating and identifying what they could change and what they would be willing to have remain unchanged. Essentially, the underlying *realization* behind all of this is, again, to demonstrate a fluid acceptance about the state of conditions taking place in one's environment. Eventually a PCL can use the term "accept" and "reject" again, since these words qualify to mean the same, though they often carry heavier emotional encoding.

---

266 **activating event** : an incident or occurrence that automatically stimulates a conscious or unrecognized reminder or 'ping' from an earlier *imprinting incident* recorded on one's own personal timeline as an emotionally charged and encoded memory; an incident or instance when thought systems are activated to determine the consequence or significance of an activity, motion or event—often demonstrated as *Activating Event → Belief Systems → Consideration.*

* For some Seekers, "leave you *unchanged*" does not register as well as "allow you to *remain*" or "keep the *same.*" For the sake of processing, the wording should be brought to acceptance as "unchanged" as quickly as possible.

Circuit-1:	Get a sense of you changing X.
	Get a sense of not changing X.
Circuit-2:	Get a sense of X. changing you.
	Get a sense of X. not changing you.
Circuit-3:	Get a sense of others changing X.
	Get a sense of others not changing X.
Circuit-1:	What do you want changed about X?
	What do you want to remain about X?
Circuit-2:	What does X. want changed about you?
	What does X. want to remain about you?
Circuit-3:	What does X. want changed about *others*?[‡]
	What does X. want to remain about *others*?

By all means a person can decide to change something, but on their own determinism and not simply because they are compelled to or even forced to by other-determined factors. In fact, the willingness to change is what empowers one to handle opposition, even if in the end we are still achieving the same result we originally intended. Likewise, the willingness for things to remain as they are also puts us in a position of acceptance and the ability to knowingly create, copy or duplicate something, again entirely on one's own determinism and without the obsessive need to repeat an action. These are some of the most critical skills of proper energy handling that seem to have escaped most mystical esoteric schools in the past.

---

‡ Alternatively, "What do others want changed about X?" or else "Others" can be treated as a "Sphere of Existence" terminal. The additional A.T. Circuit-0 would include Self as X.

:: ⁹x ::

MASTERS OF THE UNIVERSE
HELPING LIFE HELP ALL LIFE
[Extended Course]

This is a transcript of a lecture given by Joshua Free on the evening of June 21, 2020. Content of this lecture introduces course material presented by the NexGen Systemological Society (NSS) from the International School of Systemology (ISS). It is included within this volume to supplement previous chapter-lessons and for the benefit of Seekers (and Pilots) interested in experimenting with practical applications of Mardukite Zuism and Systemology (ZU)-Tech alluded to throughout this volume.

"Existence in the Physical Universe is playful. But there is no necessity for it. It isn't going anywhere. It does not have some destination that it ought to arrive at... But it is best understood as an analogy to music; because music as an art-form is essentially playful. One doesn't make the end of the composition the *point* of the composition. We don't see this expressed as something brought by our education into everyday conduct. We've got a system of schooling in society that is all graded; and what we do, is we put the child in the corridor with this grade system with a kind of 'come on kitty, kitty, kitty'—and now you go to kindergarten, and that's a great things, because when you finish that, you get into first grade. And then, 'come on,' first grade leads to second grade, and then you get out of the grade school and you go to high school—and it's revving up, the *Thing is coming!*—And you go to college, and by Jove, you get into graduate school, and when you get through with graduate school you get to go out and join the world; And then you'll get into some racket where you're selling insurance or something, and they've got that quota to make, and you're gonna make that. And all that time, that *Thing is coming!—it's coming! It's coming! That Great Thing!* The success you're working for. Then when you wake up one day, about forty years old, you say, 'My god! I've arrived! I'm there!' And you don't feel very different from what you always felt. By expectation, look at the people who live to retire and put those saving away. And then when they're sixty-five, they don't have any energy left; they're more or less impotent and uh, they go and rot in an old people's 'senior citizen's community.' Because we've simply cheated ourselves the whole way down the line. We thought of life by analogy with a 'journey' or with a 'pilgrimage', which had a serious purpose at the end, and the thing was to get to the end—success, or whatever it is, or maybe 'Heaven' after you're dead. But, we missed the point the whole way along. It was a musical thing, and you were supposed to sing or dance while the music was being played."

—Alan Watts (*1915-1973*), British Philosopher

Here we are. I always like to open these evenings with some insightful old-school quote—makes me look smarter as I'm asked to come up here to present the latest material of our

Systemology. As some of you know, I cull together a few individuals when we are about to put a new manual together and get some of the kinks on its communication worked out. I suppose in the future, it might make more sense to present a conference or workshop on the material that has already been published; and we might consider that. Only a few weeks ago, we released *"Liber-2D"* as *"Command of the Mind-Body Connection,"* which wraps up the extensive *Piloting* course that began in *"Liber-2C"* or *"Communication and Control of Energy & Power"* just a few weeks previous to the other. So, we are certainly developing *Grade-IV*, at least at a comparable rate to the solidification of *Grade-III* at the end of 2019, and no I don't count the near decade worth of underground research and development that preceded the release of *"Tablets of Destiny"* and *"Crystal Clear."*

Tonight, I basically want to reach an endpoint on the methodology presented last week,[*] to give an idea of what we are leading to at the point on the *Pathway* that is covered in this *"Liber-3C."* There is obviously a goal in mind with the way that systematic processing was presented and then carried over from *"Crystal Clear"* and *Grade-IV Professional Piloting Procedure* and in the establishment of what we expect at the finality of *Grade-IV*; the minimum conditions that we would consider an individual to have *realized* in order to say, "yep, they're at Wizard Level." Of course we are calling *Grade-IV* a "Wizard Grade" to keep any former "Masters" from being discouraged [*laughs*], but it is, in actuality, "Wizard Level-0." The end result of *Grade-IV* is the minimum that we want to be working with when confronting any further upper-level work.

The *Pathway* through the *Gateways* to higher *realizations* and points of acceptable *beingness* has been veiled—and we know this; but we have only just realized that it is *we*, each one of us, that has participated in the "covering" and "disguise" and "blocks" that bar the way out. The average standard-issue individual has not yet reached a point of *Awareness* to where they are able to confront such truths as reality; and thus they remain hidden as a Mystery—and we all know how much fear-imprinting there is on Mystery—and this is where the occupation of attention is directed: *down* into the state of unknowns that represent all that is dangerous to our survival. I suppose its actually a bit of *looking up*, if an individual as a spiritual being really were in that low of a vantage[267] point, right? Down in the sub-ZU terrain on the standard model—that's the underworld. And why anyone would want to dwell in those depths is beyond me, but considerations and actual *Awareness* really do descend that low, almost into a hiding or a non-existence. *Stoop not down...*

This is all considered by a matter of choice, no matter how much responsibility has been given up for those choices and the power of choice is resumed only with the resumption of Awareness. So, what is it that really freed the magician on Earth and the priests and priestesses and the wizards from their ties to the material world? What was it that drove the shamans and mystics and healers—before they organized into medical institutions—along

---

[*]   Referring to content presented in former lesson-chapters of this manual.
267 **vantage** : a point, place or position that offers a good view.

their *Pathway* in such a manner as to remain clear of the bombardment of fragmentation that undoubtedly surrounded them as much in their own times as we find among us today? Given that we had already had the properties of the *first, second* and *third* "Gates" to examine, it seemed that the "push-button" nature on the *fourth* should be clear to see.

Although there were many clues provided throughout *Grade-III*, the real answer was only validated by asking ourselves a series of questions regarding what we *knew* for certain of the *Pathway* that brought us here. In point: *what* was an individual doing regardless of the point on the *Pathway* they were on that was propelling them forward. Of course, we knew the small answer to that for beta-existence was simply to exist. The being is, at its lowest point of certainty, a *being*, they have some quality of *beingness* whatever POV that might be granted toward. Great, so we know we are existing; that's something. As an existence we say that an individual must be *doing* something; there is some indication of a Will or Intention that is directing the *doing* when *Self-determined.* Of course the command of *doing* falls to lower and lower control centers as the consideration of *Beingness* and a decline of *Awareness* falls. And we know that is not promoting an continued existence.

Given all of the faculties and the abilities still accessible to the Alpha Spirit—who has merely given up the regard of these things in favor of hiding in a genetic vehicle POV—it is logical that the *being*, when defragmented, would be acting and *doing* in the direction of optimum existence and increase the quality of personal ability. We *know* that we are a projection of *Spiritual "Life" Awareness* or *Zu* entangled with the Physical Universe; but to call it a *Game* and be done with it, tells us just about as much about the purposes and rules as simply introducing to a person for the first time that *"mancala is a game"* and then giving them no further information on the subject. This only proves my point to those of you that don't already know what *"mancala"* is; but it's a game from Africa that is in some ways similar, though more primitive, to the game of "Twenty-Squares" played in Babylon.

Really, any hope of producing ideal conditions, or a certainty of continuance of existence, regards an individual actually believing in a future and their own determinism to *do.* Those who have succumbed to a nihilistic approach where "everything is pointless" and "there is no future"—these people have no "hope"; there is nothing to demonstrate to them—in their reality, meaning what they are able to confront as agreeable or acceptable—that there is any possibility of change and that nothing can help anyone. *There* is the underlying fragmentation that deteriorates the individual, the home and the society. A decline in the belief, acceptance, and therefore, ability to "help" anything, anyone, anywhere. *That* is what has entrapped us in *this* Physical Universe.

Therefore, the sum all of what we are reaching for at this stage of *Grade-IV,* is defragmentation on all energetic channels regarding the subject of "help." This is what is going to open up the *Gateways to Infinity* for the Wizard-levels of our Systemology. It is a critical and pivotal point for the Seeker; which has been slowly worked up to since its *realization* during the development of *"Crystal Clear."* Before seeing it in practice for yourself, you may be wondering how this standard relates to what is presented previously in Systemology. However, the word and concept surrounding it, seems to be the most underlying "keyword" to essentially open and close the Gates to higher realizations, because those realizations directly correspond with an increased *Awareness* regarding what one can—on each channel—be to others, accept from others, and apply to Self, *always* from a cause-point.

If one takes a close examination at the role of the magicians and priests, priestesses and mystics, shamans and all those of that type down the line, what is it that these individuals are doing in the Physical Universe? They are *helping* it to exist. Their goal is the aid, assistance and, otherwise, help of *Life*. Now, we aren't talking about misguided black magicians and evil sorcerers and such here, or the wicked witch that comes in to poison your crops for not buying a love potion; we aren't referring here to just anyone who carries the guise of *knowing*, but those that *do*. And we have found, as with the other invalidation that comes with using the Mind-System as a catalyst for *being*, that it is not the "help" and "change" and "problem-solving" in itself that causes the fragmentation through encoded imprinting or programming, it is the "failure" of such efforts that hangs us up.

If you think of the individuals in your life that you have been most angry with, I can almost guarantee it is those categorized as "failed to help"—and above all of these are those persons and situations where *you've* been the one that tried to help and registered the effort as a failure; you believed you could help someone as a cause-point and failed to manifest the effect. If we consider the aid, help and assistance of *Life* at one end of a spectrum, such as on our Standard Model, and them destroy down here at zero, one can almost create a little flip-book scenario that demonstrates the lower and lower considerations for existence as one loses their *Actualized Awareness* as well; and some of this can be linked directly to the Spheres of Existence, which is how we modeled our version of Utilitarian Ethics for Mardukite Zuism.

As the whole matter of "help" and "ability" is betrayed at each Sphere of Existence, the individual finds less and less reach and acceptance in this area—which is, of course, tied to communication and the barriers and blocks set up to define the parameters of reality along the way. So let's take a minute just to track this little turn of the *Pathway* presented in *"Liber-3C"* and how it relates to what all has led us here.

So, now we have this subject of "assistance" and "help" to the greatest reach on the Standard Model, because it is the best channel of reach that pervades through each of the Gates and what we have modeled as Spheres of Existence to superimpose on our model. The Spheres, the Universes, the ZU-line; it all interrelates—as is mainly the subject of our Systemology at its core, especially if one reviews the *Grade-III* material. So let's backtrack from where we are: here is our whole world closing up on us on the failure to help and assist; which validates what we discussed regarding the fragmentation of "change" and then also the existence of "problems."

This whole cluster of stuff we call "problems" puts us out of communication and creates an automatic network of circuitry to handle the communication of reality for us, which failing to take responsibility for this, we lose the ability to clearly recall memory without fragmented imprinting and encoding; this last point being about where we left things with *Grade-III* as "Mardukite Systemology."

For *Grade-IV*, we are approaching a new vista for *Homo Novus* that we are calling "Metahuman Systemology," and by following these basic steps we have prescribed, I don't see any reason why a willing *Grade-IV* Seeker could not reach the plateau we have set up as the culmination of *Grade-IV*, assuming of course they have already worked successfully through *Grade-III* on their own, or have had a very "helpful" *Pilot* to "assist" them along the way. And now, maybe you see a hidden side to why we *Pilot*, too—or emphasize the applied spiritual technology accessible to a *Minister* that applies
this Systemology to Mardukite Zuism.

△ △ △ △ △ △ △

A Seeker can spend a great deal of time on their own "communication lagging" through all of the Self-Processing in "*Crystal Clear*," and some of the PCLs given in the *Grade-IV* "Pilot Course" and, of course all throughout the manuals we are preparing. That is certainly one option. Alternatively, a Seeker might decide to get together with a friend—presumably another *Seeker*—and proceed to work together to move through the PCLs. Now, if you don't know about how we are gauging the effectiveness of the process or what a communication lag is, this cooperative game with a friend might start to run into trouble around the point of systematic processing that we are reaching now. Of course, there is no actual reason you couldn't work through *Grade-IV* on your own, but we have found that it really helps to work with a *Pilot* that knows that they're doing.

In the past, particularly in *Grade-III*, the methodology was basic enough to where you could still get a lot of great results on your own, even if it took you a while to get through it. Even when we started to introduce basic communication training into our procedures in *Grade-IV*, the terminals and channels selected would still provide a lot of room for personal development and skill in delivering PCLs and practicing the fundamentals of what was presented a few months ago as a new *Systemology Operating Procedure* for *Grade-IV Professional Piloting*. We want Seekers to get this right because there is no point in my going on to develop a presentation of *Grade-V* until we can do this. As we have narrowed down our semantics of energy flow as a "communication line," there is very little, if anything, that can substitute this foundation for operating our spiritual technology. When you are getting right down into the core of systematic processing, there is no substitute for skilled communication.

Our emphasis in this conference has been on the problems of the Human Condition and what we can do to help the situation. That's all. [*laughs*] But these are two elements that are critical to making progress using systematic processing. Both require that the *Seeker* provide true "presence" of Awareness to the systematic processing. The ability to focus and participate in a session is characterized by these two things: ability to handle the attention that is fixed on a problem and the certainty that help is possible. We don't even have to *do* anything about it right now, we just want the Seeker *willing* to be *helped* and able to *help others*. It may seem trite to some of you, but this runs parallel with *Actualized Awareness*.

The interesting experience that one receives by achieving higher *realizations* about protest, problems, change and help—the greater we defragment these channels and increase *Awareness*, the less "driven" an individual feels in *needing* to *do* something about something. Now, what I mean is not a passive laziness; I'm speaking about the emotional urgency and mental fixations that get attached to these "trouble areas" in the game of Life. Those imprinted compelling, compulsive, turbulent hurricanes of confusions—those are states of fragmentation; they are not necessary POV.

Let's say you are *Piloting* a *Seeker* and getting a systematic session started and so you ask them if they are having any troubles or worries that are occupying their mind. Sometimes they are not very forthcoming about where their mental attentions are, which is why working at communication is so important. Or, let's say they are disconnected from everything and just *passe* with it all, because nothing bothers them, because nothing matters or something. Yeah, watch out for that guy—because he's totally out of communication with existence. But we are talking about the stuff going on in the Seeker's life that is occupying them.

For example, the Seeker you are about to process, they got a traffic ticket—a DUI or something—the weekend before. So, they are talking about the fines they have to pay, points on their ticket, the court appearance they have to show up for in two months, the affect that losing their license may have on their job...all of that. Well, *is* that problem there in the room with you? Are they coming in to arrest him? Is this session making him late for the court appointment in 60 days? Well, I can tell you that for this Seeker it is. They aren't there systematically running processing from *Self*; they are still stuck back on the weekend getting the ticket and then running through all the possible imagery that could eventually happen down the road. They are everywhere else *but* sitting right there in the chair in front of you being processed. We emphasize this *"Presence in Space-Time"* previously in our *Grade-IV* material, and it is listed as a preliminary of *SOP-2C* for a reason.

Most of the problems that we are approaching in a systematic session are not meant to be solved in session. This is a big misconception that follows along with the programming we have regarding the magicians and priestesses and healers and such in our memory. We aren't *solving* what the problem is in session; we are *solving* getting a Seeker out from the clutches of the problem. Funny thing is, when we do that, we realize that the consideration of a "problem" was just that. We want to take the hold it has off of us or get out from under it, however you want to see that. It is a POV that is held in suspension and nothing more. Okay, maybe there *are* certain actions that could be done for it, but the individual is so busy running around in their Mind feeling that they *must do* something that the door is left wide open for all that emotional imprinting to come rushing in. This is what actually keeps the Seeker suspended in the problem.

Other traditions; they have these "psycholo-spiritual advisers" or what-have-you, and you are supposed to go to them for your all your answers because they are the only one in authority within the paradigm to tell you how it is. Well, that's just nuts. Anyone that has been working with *us* over the past year knows darn well that we haven't been training *Pilots* to provide answers; we've been training them to ask questions. This is something we know took place in ancient Mesopotamia, but seemed to be popularized, or at least carried down in visible history, from the ancient Greeks—presumably attributed to Socrates, but as we know, the Classical-period civilizations had imported their philosophies, "kabbalahs," "world-trees" and Hermetic pursuits from even more ancient, nearly prehistoric, civilizations in the regions scholars once referred to as the *Ancient Near East*, which is to say Egypt and Babylon.

I tell you though: one very actualized guy that seemed to have figured this out a couple thousand years ago would go around doing wonderful things by asking a Seeker, "Do you believe I can heal you?"—"Do you *believe help is* possible?" Never said, "Okay, now I will heal you whether you like it or not." Nope. He said, "By *your own faith*, you are healed."—"You did this to your self, *buddy*; I'm just reminding you." And its funny, the authorities of the societal systems at the time didn't seem to much care for this and then somehow they got humanity to fight one another in his name for so long that any usefulness of the messages became so lost on people so long ago that now the whole system is just used to trap and control parishioners into supporting a material organization on the hope that they can buy their way into some lofty afterlife. We're going to take up the subject of *past-lives* in the higher Wizard Grades, but its pointless to drop that a Seeker at this point when they are still rustling around with taking the responsibility for—and command of—*this* one.

It is easy to see and correlate, just based on the events and handling of energy and communication in *this* life, just how the condensation of higher universes could have taken place without even looking to *divine* some new factual data and details. There is a progressive pattern that unfolds in all systems and this understanding contributes greatly to why we call our work "Systemology." It is reduced, in the past, as "general systematology"[268] in the realm of academics, but rather than using that information to work out a method of producing a *metahuman* state via spiritual technology, the material emphasis of the new millennium and its 20th century predecessor was all aimed at the external technologies and a *transhuman* evolution that would undeniably eliminate *Human Life*, or rather seal it into an even lower

---

268 **general systemology ("systematology")** : a methodology of analysis and evaluation regarding the systems —their design and function; organizing systems of interrelated information-processing in order to perform a given function or pattern of functions.

condensation of *reality* in a digital universe—which, I might add, is *beneath* this one, not some grandiose astral realm above it.

The main subject of our new *Liber-3C* manual is "universes"—specifically *this* one, and how you got your POV stuck here. This seems wildly metaphysical, but it is a fundamental ledge[269] for our (*Grade-IV*) *Metahuman Systemology*. Now we have a better understanding of the restrictions of consideration that led to the condensation of universes. We've figured upon a few of the hot-button issues; and dedicated the past six months of research and discovery toward this *Grade* of the *Pathway*. And it has been critical for demonstrating that these levels or *Grades* or *Gates* actually do mean something in regards to the Human Condition, the Alpha Spirit and the layers of miscommunication and fragmentation that have sealed in each of the higher universes and veiled them from sight. As we increase our responsibility and command for these conditions, so too will the veils be lifted.

We've come right down to it now: the final *realizations* of this step. It is the uncovering of the secret of universes; no secret really—more of a distant foggy memory that seems to come into clearer view the more we clean those lenses of perception. When we use the Standard Model to represent the Spheres of Existence—and essentially "spheres of reach and influence"—we are demonstrating the solidification and condensation of all *Awareness* and considerations of *beingness* at each stage. What we are left with is an individuated Alpha Spirit, which after a progression through various stages of betrayal in higher universes, succumbs to a very isolated existence as the standard-issue POV of humanity. You can literally gauge an individual's true chronic state of *beta-Awareness* on our scales, simply by examining the degree to which they can exchange fluid communication; and the apex of this is specifically "assistance and help."

There are some in the past that have put a lot of stock in this word "*faith.*" Well, "faith" is funny, because on a practical level it is the actualized energy generated by the being themselves. Some religious leaders seem to talk about "faith" as if its something you can *have*, in fact you had better *have* it or their whole paradigm crumbles to pieces. Where others have indicated some special rising level of "faith" as paramount,[270] we have instead found the more effective semantic as "help." This is because the degree of universal fragmentation that an individual would otherwise be told to categorize as "faith" (regarding various aspects of existence) is really a matter of where they still maintain trust, open lines of communication and a fluidity of help and assistance. When this clarity is betrayed or broken in any way, the individual goes out of communication on that line and a veil or barrier is put in its place. In many instances, the veil is a thin screen or filter that simply processes that channel of information on an automated-response mechanism. And it would be just fine to screen our

---

269 **knowledge** : clear personal processing of informed understanding; information (data) that is actualized as effectively workable understanding; a demonstrable understanding on which we may 'set' our *Awareness*—or literally a "know-ledge."

270 **paramount** : the most important; "above all else."

calls, as the Alpha Spirit first started to do in the beginning, but now we aren't even getting our messages! So, it's a slippery slope when we start to create patterns.

The Self has always considered itself to occupy some level of existential codependency with an environment, even within the highest Alpha spiritual universes—when *Self* was first individuated away from the Infinite Beingness, or Nothingness, however you are still chewing on that one. *It's a Nothingness, by the way.* But that is the Infinite Nothingness—and that is not even where the Alpha Spirit dwells as a near-static point at the edge of Infinity. But in the Alpha spiritual universes, the Alpha Spirit *is* an individuated being; they are a true individual—the "identity" part gets taken on later and is later replaced with a personality that is believed to somehow substitute individuality within the material existence. It makes it easier to anchor the Alpha Spirit to a locatable body maybe. In either case, we are systematically releasing the hold that this artificial personality has over the viewpoints, filters and mental processing of the Seeker.

"Betrayal" is probably the most severe emotional encoding that exists on a spiritual timeline. It is linked directly to this concept of "failure to help." It is the betrayal at each level of existence—or at least the perception of the same—that sealed up the Gates behind us on or descent through the universes. It is for this reason that at *Grade-IV*, we are repairing these fractured lines of communication and lifting the veil of the *fourth Gate* with a defragmentation on the subject of Help, and for the moment, we will settle for beta-defragmentation up to the Fifth Sphere of Existence, which is essentially *All Life* on Earth, all the way to include the Green World or Kingdom of Animals and Nature. The Druids, at the apex of their Mystery School, would have presumably made it at least that far—and thereafter one saint, named Francis, seems to have as well.

What we are getting down to in *Grade-IV* is a return of responsibility and control of a Seeker's considerations, which have otherwise become fixed and limiting by all that supposedly useful "experience" that they are carrying around with them; all those energetic masses of emotional imprinting and reactive-response filter-screens that can basically do all the *looking* and *reaching* on a patterned deterioration of what is considered safe and acceptable. Obviously, when a person dies they must have decided there is no where else that is safe and acceptable from that POV and that the energetic lines have all been cemented to a point of weight and mass that the person just isn't going to get up and move anymore. They come to a point where they say, you know what, I'm done.

Now, don't get me a wrong. The *Self* should be free to leave the POV of a physical body on its own determinism, particularly if it has somewhere to go, and might just as easily come back to command that organism again. This is entirely different than the desperation and confusion that results in an individual getting "spun" or feeling quasi-suicidal. You've got most of the population that doesn't know where they came from or where they're going, but they know damn sure that they can't confront any of that or even where they are right now.

I mean: you're telling me you don't know what comes next but you're in a hurry to get there? Wow!

Those individuals that didn't break the chains of *beta-existence* will wind up right back here again, or worse, since they're going to have one more layer of validated fragmentation to carry around from the start. And this is part of the decline of spiritual ability, the descent and limitation of the Alpha Spirit's POV to the human condition and naturally the condensation of the Physical Universe. Here, everything has been reduced to a near zero-point of inert matter and energy—which is still in motion, by the way, but not nearly to the same degree as what we see in the creations of the higher Spiritual or "Alpha" universes. And there are many universes; each with its own qualifications of the Gates and veils—and this is why we suggest very strongly that those esoteric practitioners previously accessing their perception of "Seven Gates" while occupying the POV from the Earth Gate (and the fragmented identity installed) have only accessed a sequence of awakening that still resides within the *First Gate*. Those that have worked so diligently with that rigid ceremonial conception of the lore may have come to a new point of *realization* at the end, but still have only pierced the first level using those methods. We've been able to reach much higher now, that is for sure.

△ △ △ △ △ △ △

I think "Help" was once recently a hot topic in some circles—back in the 1960's and 1970's was it?—something like, "Helping You Help Me Help Others Buy More Self-Help Books" or some such, wasn't it? I guess that seemed more ethical than the "How to Influence Friends and Win People" angle. But we're hitting the subject of "help and assistance" in our Systemology and I know we are in the right direction, because just the other day I heard someone go on one of the most fragmentary rants on the topic, about how no one can help anyone and how all help has strings attached and well, they don't think they can be of help to people anymore—and I think I literally heard the word "help" a dozen times in a minute or two... yes, this is a "magic button" and should not be undervalued for your ascent on the *Pathway*.

If you consider what actions we are actually taking in Mardukite Zuism & Systemology—what we are *doing*; what you are *doing*; what everyone is here today to do—we are helping others, they are helping us, they are helping help each other and ultimately, we are helping ourselves. How we do this best? Of course, we are clearing debris out of our communication channels, extending our reach and increasing our *Awareness*—and when we consider any gradient scale or chart for this, the most objective conceptual understanding we can apply on a practical level is our ability to help and assist one another, all life and ourselves.

So let's break this down; I've given you a lot of theory and examples, but let's see how this all plays out in systematic processing. We have our Standard Model and its various external spheres of existence that we correlate *Awareness* to while operating in beta. So, yeah on the one hand we have "1" as the low points of a physical body and "2" as the reactive-response mechanisms and so forth. That's the Zu-line or personal identity continuum. Superimposed on this information, we discover that the Spheres of Existence line up quite well; so we have Self as "1" and our domestic situation or home equilibrium as "2" and then our societies at "3" and on up. Those are the Spheres of Existence and it seems as though the ability and willingness to reach along these spheres can be easily defragmented using the concept of "help."

To engage or receive help from any sphere is to be in communication with it and to be willing to freely give and receive help—which is a high level of communication—on those channels. We could just as plainly state that the resolution of beta-fragmentation regards the handling of communication, problems, protest, change and willingness—and the common point that employs all of these is snow-capped with this concept of "help and assistance." Therefore, following this logical succession of techniques in *Grade-IV*, we are essentially leading to the highest point we can really understand the channels of communication in a practical way for this beta-existence and that is the assistance and help of all life to help all life to the highest regard of life as a sphere of existence, and this not only applicable to the *Gateways* of our Systemology, but has become the basic standard behind the Utilitarian Ethics of our religio-spiritual presentation of Mardukite Zuism. You see, it's not just some standard of morals of dogma like you find elsewhere; our ethics in regard to a Utilitarian viewpoint of the Standard Model actually makes sense.

We have spoken of "betrayal," and previously of failed efforts, failed communications, failed change and other control mechanisms put in place in society whereby we have become entirely fragmented on the channels of communication regarding help and assistance. And again, it is not as if someone *must* run around the world saving everyone and everything, because that would be driven some underlying implanted compulsive need. All we can determine with certainty at this juncture is that as the highest form of communication expression in *beta-existence* that we *can* act, do, know and observe upon, this button called "help" is the key to reaching beta-defragmentation, even undoing all these other short-circuited facets of life, because once we have resolved the ability and willingness to help along any channel, we know we have remedied the other forms of potential communication on that line too.

When we talk about "failed help," we do not just mean a failure *to* help, or apply the intention to *help* because that is what happens after the fact once we start to put up the communication barriers and other filter-screen mechanisms to handle all that for us. What prompts this pattern of activity is when we did make those intended attempts in the past and the efforts failed; or when others had genuinely intended to help us but it seemed that it didn't work out. These "experiences" start to get added up and associated with the concept of help, just as much as any other imprinting or programming—but this one seems more critical to selectively defragment due to the sheer amount of weight that it is given in the

Mind, concerning personal ability, willingness and reach. Since the Mind has already associated our ability to help with our ability to reach, it makes sense to work with this as part of systematic processing as opposed to simply protest its truth.

Considering the types of issues that have been approached in *Grade-IV* systematic processing —communications, problems, changes, protests, willingness, reach, acceptability, conflict and the rest—it is only logical that we should approach the subject of "help" as a hot-button for achieving spiritual evolution as *Homo Novus*, and at the very least extend the reach of our metahuman Systemology. In theory, this is already dangerous territory, because this is the power to end wars; and certainly there are more than a few authorities in this realm that benefit from keeping the masses suspended in confusion. But just imagine if both sides of conflict were to focus on how they could help an enemy and how their enemy could help them; well, help them other than being dead of course. I'm meaning that we apply the same methods of achieving defragmented fluidity as you've seen (with the previous lesson-chapters). But you'll sometimes get that kind of response early on or in low-level states with systematic help processing. The *Pilot* asks, well, "How could so-and-so help you?" And the guy's all like, "well, the sonofabitch could be dead, not exist, that'd be a big help to me..." You know? And a well-trained *Pilot*, maintaining their composure all professional-like, not phased, will just acknowledge an acceptance of the communication—I mean, the guy gave an answer, right?—and we ask again.

Because *Grade-IV Professional Pilots* are little more further along than the instruction given in *Grade-III* material, we don't necessarily emphasize a lot of fluid two-way communication in some of these systematic processes, not because they wouldn't be more effective if you did— because they would—but unfortunately its just a skill that hasn't been mastered yet by many (of you) that are a part of our underground research and practical experimentation. Thus, we have emphasized repetition or processing that operates with repetitive PCLs that lead one toward greater *realizations* the more they work with it—but this can certainly be assisted with proper handling of communication, which we expressed in the *Grade-IV Pilot Training* in *Liber-2C* and *2D*. Just don't confuse it with what you are used to when having a casual conversation or shooting the breeze with someone. And by no means does a *Pilot* offer direct advice as to what a *Seeker* should *do* about anything. Don't start telling your *Seekers* to leave their wives and husbands and stuff or you'll end up the one in the stew.

Δ Δ Δ Δ Δ Δ Δ

When we expand our processing to include the general "terminals" of the Mind-System as opposed to specific events and incidents, we approach a wider range of potential *recall* in our processing. This is what immediately led to our interest in exploring "past lives" in future Wizard Grades, because it seems to be about the only way we can be certain of approaching anything close to a standard operating procedure for *Alpha defragmentation*, but I'm not trying to get you focused on that direction yet; simply the consideration that, for example, the treatment of "parents" or "bosses" as a terminal for systematic processing accesses more of the operating system that only treating a specific target from this life, such as your biological or adoptive parent from this life.

By tapping into the greater content of a "terminal" with a broad approach, there are reactive-response tendencies and programming that may resurface[271] that just don't seem to fit with events and memories of this lifetime—and this is when we realize we are encroaching on new territory for Systemology that I intend, as I said, to take up in *Grade-V*. But, let's just get a handle on things that we have here before us in taking responsibility for, not necessarily everything that happened *to* us in this lifetime; but the mental imagery and storage and reactive-response encoding and associations that we are carrying around as a result of it. This state of responsibility alone would border on true *metahumanism*. But it should be understood that with each lifetime we have carried more and more of the same "game" with us and are undoubtedly it playing out over and over in the roles we take on and the personality-mechanisms attached to it.

The PCLs for *Grade-IV* systematic help processing use the same circuit patterns and methodology as what you learned for "*Route-3*"[*] and "*Route-2*."[‡] We are still using the same procedures to defragment the channels; I'm not sure if you were expecting some radical new curve-ball here at the end, but no this is really basic if you've been following along to this point. Naturally, I want each and every one of you to be successful at this, which is why I've dedicated my life to breaking it down this way for you and doing the grunt work and long hours in the library and the thousands upon thousands of dollars in research expenses to bring this forward. Most of you that have been following along over the years know that I've spent a lot of time in the underground figuring on these *Gates,* and ways of accessing them for real—and while Systemology might seem like a lot of fancy word-play and creative psychology, I can say that after a quarter-of-a-century journey of being a messenger on these matters, I have found no better access point to achieve the genuine goals behind all former traditions of spirituality, mysticism, religion and philosophy. This is a delivery of the "Great Work" in its purest simplicity—because anything more esoteric and you could miss it. And we've already been coming back here to Earth over and over for however many thousands of years; always missing it. So let's not miss it this time.

Unless a Seeker or Pilot is using a *Recall* approach to events and instances—which is what you would do to release emotional turbulence encountered—the "Route-3" approach often allows for more creative answers that are not restricted to a specific time something happened. You

---

271 **resurface** : to return to, or bring up to, the "surface" what has been submerged; in *NexGen Systemology*— relating specifically to processes where a *Seeker* recalls blocked energy stored covertly as emotional "imprints" (by the RCC) so that it may be effectively defragmented from the "*ZU-line*" (by the MCC).

* "*Communication and Control of Energy & Power*" (*Liber-2C*) and "*Command of the Mind-Body Connection*" (*Liber-2D*).

‡ "*Crystal Clear*" (*Liber-2B*).

will notice this in the *Grade-III* and *Grade-IV* PCLs that use the word *"could"* or *"would"* rather than "recall a time when..." We see an expansion on this later in *Liber-3D,* a forthcoming title for *Grade-IV* that should just about get us wrapped up on the present ledge of goals, where the other *"Route"* introduced in *SOP-2C* speaks of "Imagining" and "Creating" and "Inventing"—which is a clue I'm giving you that we will be very soon dealing with a handling of "mental imagery" directly.

When we use a PCL that says *"could,"* we are in no way implying or insisting an action—only the consideration. An individual doesn't necessarily even have to help their enemies, but the free consideration that they *could* prevents the hatred from swelling up into an emotional mass that shuts down energetic communication along that channel and setting up a filter-screen of automatic response-reaction. If all you do is spend all day worrying about how to destroy your enemies, I can guarantee you that the one that will get messed up in the end is *you.* It isn't that you can't defend yourself in some situation either, that isn't what I'm talking about. But keep in mind that most of what we like to think we are in control of regarding our thoughts and what we think we are in command of concerning our actions is filtered through many filters and slams around many masses and barriers of failure and error, pain and loss, hatred and jealousy and so forth. If a person is having to duck-and-go around all of that, they are not in command of the Mind-Body connection, no matter how much they think they have got it all figured. It took me a long time to realize that "Turn the other cheek" simply meant "don't react."

It should be noticed—if you've been getting any actual progression on the *Pathway* from a genuine application of this work—that you've been coming up against layers of resistance as you've been playing at the game of systematic processing since *Grade-III.* Each time we apply these methods, or run a series of PCLs that are meant to increase our considerations, there is a series of barriers that one is breaking through with each level of responses. If you aren't hitting these barriers, you may not be running the processing long enough. Usually the first time you are running low on answers, you are not running out of answers, you've just cleared a level of rubble that you've been working with and if you get past the next veil or layer of resistance, you will find a whole new level of stuff to work with, no matter how ridiculous it seems for this *beta-existence,* those answers are a new level of consideration. This is generally only worked until a *Seeker* no longer carries a heavy mass or emotional charge on that line, but in theory, you could extend the basic methodology of systematic processing all the way up the chain of existence to the highest sphere fathomable.

If you want to apply *Route-2 AR* in order to get a session moving in the direction of contacting the help channels, then by all means. That application should, however, be run on the circuits of *"Route-3"* and preferably with full two-way communication about the subject matter, making certain that the Seeker is willing to communicate on these lines with the *Pilot.* All you would do here is run them through the circuits on recall with: them helping another; another helping them; another helping others or another; and of course self helping self, if you are running all the way to *A.T.* with it. If we are being more general with the terminals, however, we can simply apply a PCL directly to the concept of willingness to help on any terminal.

Circuit-1:	Who (or what) would you be willing to help?
	Who (or what) would be acceptable for you to help? (*Alternative*)
Circuit-2:	Who (or what) would you be willing to have help you?
	Who (or what) would you be willing to accept help from? (*Alternative*)
Circuit-3:	Who (or what) would you be willing to have others help?
	Who (or what) would be acceptable for others to help? (*Alternative*)

You may find it the case that you need to alter the wording of some PCLs in the beginning or early on in a *Seeker's* processing, or else be very certain they are clear on the meaning of the semantics applied in Mardukite Zuism and Systemology before they are applied to systematic processing. Of course, this is what makes increasingly higher grade processing possible: the very fact that as an individual is getting clearer and clearer on a broader range of communication, they are able to reach just a little further with that. Once we are clear on our level of realization on one ledge of knowing, then we can, with surefooted[272] certainty, reach up toward the next.

The most general and basic systematic processing we have on this lines has run in "New Thought" circles for nearly one-hundred years. The most important part of defragmentation is clearing the predisposed inclinations and associations that close off our communicative energies along the channels we are connected to. Only once these are all opened up can we hope to be free of the fixed considerations that hold the abilities of the Mind-System to *this* Physical Universe. So, we have the *Pilot* direct a five-way series of PCLs that should get a Seeker over the basic humps of help.

Circuit-1:	How could you help "someone else" (*terminal*)? {Circuit-1}
Circuit-2:	How could "someone else" (*terminal*) help you? {Circuit-2}
Circuit-3(a):	How could "someone else" (*terminal*) help others?
Circuit-3(b):	How could "someone else" (*terminal*) help themselves?
Circuit-AT:	How could you help yourself?

Keeping in mind that we like to break up a long intensive series of subjective processing with some objective processing, the same formulas may be applied to "*Bell, Book & Candle*" or other objects; even a trouble-terminal or problem object in the external universe. This is very similar to what is described earlier in *Grade-IV* regarding lines of communication with terminals. Instead of staying on that same general idea of "communicating with" such and such a terminal or sphere or object or concept, we are speaking of "helping" as an active expression of

---

272 **surefooted** : proceeding surely; not likely to stumble or fall.

high-level communication that promotes the Prime Directive of existence toward its ideal state or condition of beingness. If each and every one of us was promoting each others Prime Directive, none of us would have needed to be down here in this dungeon universe. You can work up this chain on the practice of "get the sense of" type PCLs to start loosening the fragments of a *Grade-III* Seeker too. This may be applied to anyone though, because it denotes "action" or "motion" when saying "helping" instead of "help" and therefore would be a logical progression from the former five-way series.

Circuit-1:	Get the sense of you helping "someone else" (*terminal*)? {Circuit-1}
Circuit-2:	Get the sense of "someone else" (*terminal*) helping you? {Circuit-2}
Circuit-3(a):	Get the sense of "someone else" (*terminal*) helping others?
Circuit-3(b):	Get the sense of "someone else" (*terminal*) helping themselves?
Circuit-AT:	Get the sense of you helping yourself?

It is obvious that the RCC has registered and imprinted many "failures" and "losses" onto our stores of experience and memory—but it has not always been found to be the best route of systematic defragmentation to exclusively process out turbulence with commands that focus on "failure." It is true that "failure" is a hot-button and we *should* be able to confront it without emotional reactivity, but this is not what we are demanding of our Seekers at this time; only that they work toward it with each pass through the material. But yeah, if you start running at the Seeker with PCLs on all the times everything has failed, you're just going to get them spun and shut down to where they don't want to communicate at all. This is a dangerous point to reach since all of our systematic processing is rooted in communication.

There are lower levels of processing one could scrape at if necessary. All we need to do is go back into our arsenal of PCLs rendered for this new *Liber-3C* and in the appendix of our new *Liber-2D*[*] and start plugging in help. Most of them will work. For example, you could theoretically run something as basic as "What help could you accept?" and "What help can you reject?" up to a point where the Seeker understands that they should not automatically reject any consideration of true help.

If you want to link this work with problems, you might use something like "What problem *could* your help be to another?" or even "What problem *has* your help been to another?" if you are approaching it from the *Analytical Recall* angle. If you do this, make certain to then apply the other circuits from "*Route-3*" for a full *beta-defragmentation* regimen, such as "What problem another's help has been to them" and so forth. Once you understand the patterns inherent in the PCLs of our systematic processing, it becomes easier to apply the right type or level of processing to the individual, knowing that you have an entire gradient scale to work with.

---

[*] *"Command of the Mind-Body Connection"* by Joshua Free.

When dealing with "past recall" or experiential defragmentation of imprinting and programming, it's not as if we can completely get around the energy attached to "failed help," but that isn't the wording that we plug into our PCLs. You combine *Recall* with "*Route-3*" and alternate "help" and "no help" on each circuit. That's all. Very basic. In the actual communication of it, you may need to state it in terms of "given help" and "not given help." And we say "not given help" or "not giving help" in the place of "failed help," because even when someone had said they are helping, but it turned out they were not or it did not seem to produce the proper result, that is still registered as a failed communication on the line of help, you see? The same is for those situations of betrayal by certain social roles that are supposed to help but don't and then we cut off lines of communication and put up blocks because we don't see how we can help them any longer either.

Circuit-1:	What help have you given to (another)?
	What help have you not given to (another)?
Circuit-2:	What help has (another) given to you?
	What help has (another) not given to you?
Circuit-3:	What help has (another) given others?
	What help has (another) not given to others?
Circuit-AT:	What help have you given yourself?
	What help have you not given yourself?

This general formula works pretty well with all terminals and Spheres of Existence. You can then apply an extension of the five-way formula I mentioned previously, but adding the alternations of "no help" on each circuit. And this should be done with as many "terminals" as possible, with particular attention to those representing an "trouble targets" in a Seeker's life. This should be run on any terminal that symbolizes the various Spheres of Existence and influence, primarily—let me read off a list here—mother, father, or parent, guardian; child, stranger; spouse or lover; teacher; healer or doctor; priestess, priest or even holy man; policeman, states official; and yes, I want to see *Grade-IV* Seekers working toward defragmentation on their connection with the Green World of the animal kingdom and nature as well. There's been a communication break between the Human Condition and the planet Earth as a living organism for far far too long.

Honestly, the easiest answer that I've come up with over the past several years, on the subject of help is "*processing*"—systematic processing via the methodology presented in Mardukite Zuism and its advanced *metahuman systemology*. Processing. Help? You processing others. Teaching others to process others. You getting processed.

But above all else—HELP ONE ANOTHER! Thank you!

:: APPENDIX ::

GENERAL GRADE-IV INTRODUCTION BASED ON LECTURES BY JOSHUA FREE
NEXGEN SYSTEMOLOGY PROFESSIONAL PILOTING PROCEDURES COURSE

SYSTEMOLOGY is the practical application of "systems theory"[273] to the study and spiritual experience of *Life, the Universe and Everything*. This "technology" is sometimes referred to as "NexGen Systemology," "Mardukite Systemology"[274] and/or "Spiritual Systemology" in order to distinguish it from other academic sciences and modern uses of the term.

MARDUKITE SYSTEMOLOGY is a developing product or result from intensive work conducted by an official extension of the "Mardukite Research Organization" in 2010, as directed and recorded by mystic philosopher and underground esoteric author, Joshua Free. *Systemology* work is treated as the practical spiritual philosophy and technology accessible to us today (and for the future) from the oldest cuneiform writings and Arcane Tablets set down in ancient Babylon[275]—in the heart of Mesopotamia[276] and "Sumerian"[277] civilization. (This is revealed extensively within *"Grade-III Mardukite Systemology."*)

There are also those Seekers that will treat this same material (below the knowledge "Grade" of Systemology) in an exclusively religious or mystical appreciation as MARDUKITE ZUISM— although the two are not mutually exclusive. At this time we are treating *NexGen Systemology* as the "upper level" applied spiritual philosophy, techniques for spiritual counseling and spiritual evolution (Self-Actualization) methodology *of* the former "religious" understanding of *Mardukite Zuism*, in any of its derivative forms, including "Mesopotamian Neopaganism" or "Hermetic Mysticism" or any surviving faction today that has tapped the 'stream' that leads behind the "Ancient Mystery School."

"Grade-IV Systemology" materials delivered within this "intermediate course" are presented in a straightforward and direct manner so that they may be understood by anyone—to the extent or level an individual has actualized a "ledge" of understanding for. For this reason our previous Grades of Research and Discovery, which a Seeker may study, are structured to represent three "levels" of understanding leading toward a basic state of "Self-Honesty."[278]

---

273 **general systemology ("systematology")** : a methodology of analysis and evaluation regarding the systems —their design and function; organizing systems of interrelated information-processing in order to perform a given function or pattern of functions.

274 **Mardukite Zuism** : a Mesopotamian-themed (Babylonian-oriented) religious philosophy and tradition applying the spiritual technology based on *Arcane Tablets* in combination the "Tech" from *NexGen Systemology*; first developed in the New Age underground by Joshua Free in 2008 and realized publicly in 2009 with the formal establishment of the *"Mardukite Chamberlains."*

275 **Babylonian** : the Mesopotamian civilization that evolved from *Sumer*; the inception of all societal and religious systematization.

276 **Mesopotamia** : land between Tigris and Euphrates River; modern-day Iraq.

277 **Sumerian** : ancient civilization of *Sumer*, founded in Mesopotamia c. 5000 B.C.

278 **Self-honesty** : the *alpha* state of *being* and *knowing*; clear and present total *Awareness* of-and-as *Self*, in its

The subject of each Grade is always the same: *Life, the Universe and Everything.* The only factor that differs is the "level"[279] of understanding used to treat the study and practice of information. "*Cultural*" and "*language*" themes are another factor that tends to differ across the historical timeline on Earth. We resolve many of these semantic issues in "Grade III" when working toward a standard universally applicable terminology, demonstrable as the "Standard Model[280] of Systemology"—which includes the "*ZU-line.*"[281]

The premise behind current work is provided in "*Grade-III Mardukite Systemology*" material available in core *Grade-III* textbooks—*The Tablets of Destiny* and *Crystal Clear*—both of which are collected within the Master Edition *Grade-III* anthology: *The Systemology Handbook.*

Application of "Systemological Self-Processing"[282] developed after a decade of additional esoteric experimental research privately conducted by remote members of the underground "NexGen Systemological Society."[283] This ongoing exploration into "applied spiritual philosophy" is established in light of all collected wisdom from various mystical and spiritual pursuits during the last 6,000 years of record history—most of which is found to be either erroneous and/or unworkable in effectively producing consistent results toward a higher state of Actualized Awareness.

---

most basic and true proactive expression of itself as *Spirit* or *I-AM*—free of artificial attachments, perceptive filters and other emotionally-reactive or mentally-conditioned programming imposed on the human condition by the systematized physical world.

279 **level** : a physical or conceptual *tier* (or plane) relative to a *scale* above and below it; a significant *gradient* observable as a *foundation* (or surface) built upon and subsequent to other levels of a totality or whole; a *set* of "*parameters*" with respect to other such *sets* along a *continuum*; in *NexGen Systemology*, a *Seeker's* understanding, *Awareness* as *Self* and the formal grades of material/instruction are all treated as "*levels.*"

280 **standard model** : a fundamental *structure* or symbolic construct used to evaluate a complete *set* in *continuity* relative to itself and variable to all other *dynamic systems* as graphed or calculated by *logic*; in *NexGen Systemology*—a "*monistic continuity model*" demonstrating *total system* interconnectivity "above" and "below" observation of any apparent *parameters*; the *ZU-line* represented as a singular vertical (y-axis) waveform in space across dimensional levels (universes) without charting any specific movement across a dimensional time-graph x-axis.

281 **ZU-line** : a spectrum of *Spiritual Life Energy (ZU)* as conceived on the Standard Model of Systemology; an energetic channel of *Identity-continuum* connecting the *Awareness (ZU)* of an *Alpha-Spirit* with "*Infinity*"; a *Lifeline* on which *Awareness (ZU)* extends from the direction of the "Spiritual Universe" (AN) as its *alpha state* through an entire possible range of activity in its *beta state*, experienced as a *genetic-entity* occupying the *Physical Universe (KI)*; the Standard Model demonstrates the *Zu-line* interacting with spheres of existence.

282 **processing, systematic** : the inner-workings or "through-put" result of systems; in *NexGen Systemology*, a methodology of applied spiritual technology used toward personal Self-Actualization; methods of selective directed attention, communicated language and associative imagery that targets an increase in personal control of the human condition.

283 **NexGen Systemology** : a modern tradition of applied religious philosophy and spiritual technology based on *Arcane Tablets* in combination with "*general systemology*" and "*games theory*" developed in the New Age underground by Joshua Free in 2011 as an advanced futurist extension of the "*Mardukite Chamberlains.*"

The functional purpose of "Systemology Processing"—whether performed by one's *Self* alone out of a workbook (like *Crystal Clear*), with the assistance of a friend or by a "professional systemology pilot"[284]—is for a *Seeker* to effectively <u>actualize</u>[285] true <u>realizations</u>[286] that produce positive movement on what is referred to as THE PATHWAY TO SELF-HONESTY[287]—which is also the technical designation for *Grade-III* materials otherwise referred to as "Mardukite Systemology."

During a twenty-five year engagement with the underground esoteric and "New Age" community, Joshua Free discovered that the majority of practitioners following the "Route of Magick and Mysticism"* or "Route of Druidism and Dragon Legacy"‡ (and other esoteric traditions amalgamated from diverse well known "organizations," "orders" and "fellowships") were not independently arriving at the intended *realizations* from these philosophies, much less an *actualization* of the same.

This is one of the stumbling blocks that Joshua Free discovered concerning most contemporary approaches to "enlightenment" and metaphysical "Self-Help" regimens. One primary goal of systemology—which should be evident in the systematic presentation and demonstrable "Self-Processing" techniques in *Grade III*—is to raise an individual's ("*Seeker's*") state of Actualized Awareness. This requires, by definition, bringing what is hidden into the light, or else carrying those aspects of "consciousness"[288] that exist below the level of analytical surface thought up to such levels where they may be treated "consciously" by *Self*—the true and actual *spiritual Self;* what we call the "Alpha Spirit."[289]

---

284 **pilot** : the steersman of a ship; in *NexGen Systemology*—an individual qualified to operate *Systemology Processing* for other *Seekers* on the *Pathway to Self-Honesty*.

285 **actualization** : to make actual; to bring into Reality; to realize fully in *Awareness*.

286 **realization** : the clear perception of an understanding; to make "real" or give "reality" to so as to grant a state of "beingness" or "being as it is"; the state or instance of coming to an *Awareness*; in *NexGen Systemology*, "gnosis" or true knowledge achieved during *systematic processing*; achievement of a new (or "higher") cognition, true knowledge or perception of Self in relation to reality.

287 **Self-honesty** : the *alpha* state of *being* and *knowing*; clear and present total *Awareness* of-and-as *Self*, in its most basic and true proactive expression of itself as *Spirit* or *I-AM*—free of artificial attachments, perceptive filters and other emotionally-reactive or mentally-conditioned programming imposed on the human condition by the systematized physical world.

* Treated as "*Grade-I Part-A*" by the *International School of Systemology (ISS)*.

‡ Treated as "*Grade-I Part-D*" by the *International School of Systemology (ISS)*; see "*Merlyn's Complete Book of Druidism*" *(Grade-I Master Edition Anthology)* by Joshua Free.

288 **consciousness** : the energetic flow of *Awareness*; the Principle System of *Awareness* that is spiritual in nature, which demonstrates potential interaction with all degrees of the Physical Universe; the *Beingness* component of our existence in *Spirit*; the Principle System of *Awareness* as *Spirit* that directs action in the Mind-System.

289 **alpha-spirit** : a "spiritual" *Life*-form; the "true" *Self* or I-AM; the spiritual (*alpha*) *Self* that is animating the (*beta*) physical body or "*genetic vehicle*" using a continuous *Lifeline* of spiritual ("*ZU*") energy; an individual spiritual (*alpha*) entity possessing no physical mass or measurable waveform (motion) in the Physical Universe as itself, so it animates the (*beta*) physical body or "*genetic vehicle*" as a catalyst to experience *Self*-determined causality in effect within the *Physical Universe*.

Raising a *Seeker's* "level" of *Awareness—Actualized Awareness—*means very simply bringing more of an individual's "actual present space-time" (beta) *Awareness*[290] in "phase"[291] or "synch" with the "Alpha Spirit" (the True "I-AM" Self). This brings the power and attention of *Awareness* more under the control of the *Seeker*, which is to say "clear communications" of *actual* potential.

> "It is my goal for NexGen Systemology that we can elevate the *Actualized Awareness* of all *Seekers—*all *Humans—*on Earth to provide a true vehicle for their spiritual evolution in Self-Honesty. It is the duty of every Systemology Pilot and Mardukite Minister to bring the conscious *Awareness* of all *Humanity* up and out from the murky muddy heavy sticky dross[292] that they are subjected to; all of the solidarity and barriers leading to the acceptance that we must remain confined and entrapped only to the lowest level of conceptual consideration. It is by cumulatively shedding skin and layers of everything that is not the true I-AM Self that a *Seeker* ascends through sequential Gates of Higher Understanding—meaning true realization in 'Self-Honesty' and *Actualized Awareness.*"

Grade-IV is clearly an intermediate stepping stone between two great planes of realization:

a.) what has come before—treated as the "Master" levels, including basic *Grades I-II,* and the *Grade-III* "Pathway to Self-Honesty" distinguishing "Mardukite Systemology"; and

b.) what we are now leading into—using *Grade-IV* as a launch point toward higher "Wizard" levels, distinguishing the "Actualized Technician" (A.T.) Grades still forthcoming.

It is easy to differentiate material from the "Master Grades" (*I–III*)[293] that consist of a considerable amount of reading, research and personal experimentation from the introduction of "Professional Piloting Procedure" at *Grade-IV.* At this present Grade of research and discovery, both the Pilot *and* Seeker are developing skills, earning education and strengthening personal certainty of the "Alpha Spirit" as an "Actualized Technician" of spiritual technology.

---

290 **beta (awareness)** : all consciousness activity ("*Awareness*") in the "Physical Universe" (KI) or else *beta-existence*; *Awareness* within the range of the *genetic-body,* including material thoughts, emotional responses and physical motors; personal *Awareness* of physical energy and physical matter moving through physical space and experienced as "time"; the *Awareness* held by *Self* that is restricted to a physical organic *Lifeform* or "*genetic vehicle*" in which it experiences causality in the *Physical Universe.*

291 **phase alignment** or "*in phase*" : to be in synch, in step or aligned properly with something else in order to increase the total strength value; in *NexGen Systemology*—referring to alignment of *Awareness* with a particular identity, space or time (such as being *in* Self *in* present *space* and *time*).

292 **dross** : prime material; specifically waste-matter or refuse; the discarded remains collected together.

293 **"Master Grades"** : literary materials by Joshua Free (written between 1995 and 2019) revised and compiled for the International School of Systemology instructional grades—"Route of Magick & Mysticism" (*Grade I, Part A*), "Route of Druidism & Dragon Legacy" (*Grade I, Part D*), "Route of Mesopotamian Mysteries" (Grade II) and "Route of Mardukite Systemology" (*Grade III*), collectively known as the "Pathway to Self-Honesty."

Joshua Free first announced the decision to integrate "Systematic Self-Processing" and "Professional Piloting" into Mardukite Systemology during a lecture given on August 9, 2019.[*]

"There are many solitary methods of heightening Awareness and increasing mental skills necessary for advanced *'processes,'* but I bring up this example... because when engaged in [professional] processing, there are two people involved: one of them is going through the *'processing'* toward Self-Honesty and one of them is assisting from a point of Self-Honesty. We identify the one receiving the service, or going through the *'processing,'* as the Seeker. In order to differentiate a very specific role that the assistant has in this process, the individual administering the *'processes'* is referred to as the Pilot. And let me make this point clear from the get go: the Pilot is specifically and exclusively responsible for Self-Honestly assisting the Seeker in reaching their chosen *destination*—nothing more or less. The Pilot is not a tour guide; not an interpreter; not a doctor; certainly not a therapist in the traditional sense—they are offering no actual advice toward or against anything that is uncovered as a result of systematic processing. Any and all realizations are meant for the Seeker to discover, determine and actualize on their own. The Seeker merely has the confidence now of knowing there is a safety net of travel by someone who has already been where they want to go!"

Based on this description, your first thoughts may be that this *Grade-IV* course must pertain exclusively to some rigorous "procedures" and esoteric philosophies that are useful only to those upper-level students of our underground brand of dispensed knowledge. This could not be any further from the truth. *Anyone* can benefit from the basic instruction and applied spiritual philosophies explained and demonstrated throughout this book.

Many believe that the purpose of all true spiritual, mystical and philosophical avenues or "routes" engaged as paradigms to treat the experience and interaction of *Life, the Universe and Everything* are headed in the "same direction" and are to be held in equal regard. If truth, the historical timeline and demonstrable workable effectiveness are any definitive indicators, we can be most certain that the resulting "destination" for the plethora of "routes to knowledge" brewed within the intellectual labyrinths of the "human condition" are anything but equal to one another.

In fact, what—if anything—could be truly demonstrated as "equal" in its actual properties of existence and interaction in this solidified and condensed "physical universe" (which we call *"beta existence")*[294] unless we are referring to the exact "same" thing in two spots in space?

---

[*]   Transcripts appear in the text of an extension course published in *"The Tablets of Destiny" (Liber-One)* and reprinted in the complete *Grade-III* anthology *"The Systemology Handbook."*

294 **beta (existence)** : all manifestation in the "Physical Universe" (KI); the "Physical" state of existence consisting of vibrations of physical energy and physical matter moving through physical space and experienced as "time"; the conditions of *Awareness* for the *Alpha-spirit (Self)* as a physical organic *Lifeform* or *"genetic vehicle"* in which it experiences causality in the *Physical Universe.*

Such *is* actually possible, though the "human condition"[295] tends not to experience such phenomenon at the mundane levels of "standard issue"[296] sensory reception. In fact, "associative knowledge"[297] and the inability to *distinguish* "things" in the universe is one of the primary sources of "personal fragmentation"[298]—which is one of the main subjects of our systemology.

The purpose of systemology and systematic piloting is to support responsibility of the Prime Directive of all *Life, the Universe and Everything*—which is *to exist* and to act toward a continuation of *existence*—and for this purpose: to actualize the highest reach as "cause" on the spheres, which is to say "defragmenting"[299] or clearing energetic channels all the way up to our highest states of knowing and being, or otherwise Actualized Awareness.

Being a high-level "cause" means the Self-directed communication and control of energy and power in such a manner as to consistently act toward the continuation of personal existence at is highest and truest "Alpha"[300] state. This means Self-directing effects that will promote the highest ideals of "utility" on the Zu-line and its reach across the Spheres of existence from the point-of-view (POV)[301] of *Self* here in this moment as it experiences the physical continuity of a low-level densely solidified universe as *beta-existence*.

A systemologist soon learns the fastest route toward actualizing the highest extent of reach as "cause" is to act toward assisting all existence insofar as it mutually helps to maintain the

---

295 **Human Condition** : a standard default state of Human experience.

296 **standard issue** : equally dispensed to all without consideration.

297 **associative knowledge** : significance or meaning of a facet or aspect assigned to (or considered to have) a direct relationship with another facet; to connect or relate ideas or facets of existence with one another; a reactive-response image, emotion or conception that is suggested by (or directly accompanies) something other than itself; in traditional systems logic, an equivalency of significance or meaning between facets or sets that are grouped together, such as in *(a + b) + c = a + (b + c)*; in NexGen Systemology, erroneous associative knowledge is assignment of the same value to all facets or parts considered as related (even when they are not actually so), such as in *a = a, b = a, c = a* and so forth without distinction.

298 **fragmentation** : breaking into parts and scattering the pieces; the *fractioning* of wholeness or the *fracture* of a holistic interconnected *alpha* state, favoring observational *Awareness* of perceived connectivity between parts; *discontinuity*; separation of a totality into parts; in *NexGen Systemology*, a person outside a state of *Self-Honesty* is said to be *fragmented*.

299 **defragmentation** : the *reparation* of wholeness; a process of removing "*fragmentation*" in data or knowledge to provide a clear understanding; applying techniques and processes that promote a *holistic* interconnected *alpha* state, favoring observational *Awareness* of continuity in all spiritual and physical systems; in *NexGen Systemology*, a "*Seeker*" achieving an actualized state of basic "*Self-Honest Awareness*" is said to be *defragmented*.

300 **alpha** : the first, primary, basic, superior or beginning of some form.

301 **point-of-view (POV)** : an opinion or attitude as expressed from a specific identity-phase; a specific standpoint or vantage-point; a definitive manner of consideration specific to an individual phase or identity; a place or position affording a specific view or vantage; circumstances and programming of an individual that is conducive to a particular response, consideration or belief-set (paradigm); a position (consideration) or place (location) that provides a specific view or perspective (subjective) on experience (of the objective).

Prime Directive at all levels—"help" being one of the highest forms of communication, which not only allows an individual to *be* at a position of *cause*, but also increases energy frequencies on the Zu-line (which satisfy the necessary requirements of a Prime Directive). The ability to extend our reach as "cause" is accelerated by the "help" and "assistance" of whatever we may take responsibility for, even if its only going to be a responsibility of being in communication with the universe; any universe.

When we consider the role of Pilots in Systemology—they are helping and assisting a Seeker, which in turn is helping and assisting the Pilot's reach on the Prime Directive. A Seeker must be willing *to be* helped and assisted—and be willing to help and assist the Pilot—by providing a full "attention" (presence) for participation in the session and processing communications.

Academic and practical emphasis during this first installments of *Grade IV* concerns defragmentation of <u>communication</u>, personal communication systems and control of the same (in *Liber-2C*[302] and *Liber-2D*).[303] In no short time it will become completely evident just why this is so important; particularly as a stable orientation point to launch these higher grades of instruction. With that resolved, the second core installments (*Liber-3C*[304] and *Liber-3D*)[305] prioritizes instruction and techniques that put the personal power and creative energy of the <u>imagination</u>[306] back under full determination of Self.

With tools and training provided as *Grade-IV* "*Professional Piloting Procedure,*" an interested reader, *Seeker* or *Master* may strongly benefit from attaining a stronger certainty of control over personal <u>communication</u> and <u>imagination</u> as Self-directed by Will[307] and Intention.

"And when one truly realizes the full considerations that a combination of <u>communication</u> and <u>imagination</u> truly has upon the individual, an entirely new or previously unreachable universe of possibilities suddenly becomes real again; becomes a potential Reality again within the reach of Self as Alpha Spirit. Each an every one of us is a partic-

---

302 **Liber-2C** : First published in April 2020 as "*Communication and Control of Energy & Power: The Magic of Will & Intention (Volume One)*" by Joshua Free; an integral part of the *Grade-IV* Professional Piloting Course.

303 **Liber-2D** : First published in June 2020 as "*Command of the Mind-Body Connection: The Magic of Will & Intention" (Volume Two)*" by Joshua Free; an integral part of the *Grade-IV* Professional Piloting Course.

304 **Liber-3C** : First published in July 2020 as "*Now You Know: The Truth About Universes & How You Got Stuck in One*" by Joshua Free; a manual in the *Grade-IV* Metahuman Systemology series.

305 **Liber-3D** : A forthcoming manual in the *Grade-IV* Metahuman Systemology series.

306 **imagination** : the ability to create imagery in the mind at will and change or alter it as desired; the ability to create, change and dissolve mental images on command or as an act of will; to create a mental image or have associated imagery displayed (or "conjured") in the mind that may or may not be treated as real (or memory recall) and may or may not accurately duplicate objective reality.

307 **will** *or* **WILL** : in *NexGen Systemology* (from the *Standard Model*)—the spiritual ability at (5.0) on the *Standard Model* of an *Alpha Spirit* (7.0) to apply *intention* as "Cause" from a higher order of reasoning and consideration (6.0) than the thoughts found in *beta-existence,* where it manifests as "effect" below (4.0).

ipant in the creation of universes and realities and we have the responsibility to our Self to permit the highest freedom of the Alpha Spirit to once again unfold as the present Awareness as Self. This is a state that is completely within reach of all individuals on planet Earth today; all we have to do is free ourselves to create a better world. So, let's get together and help one another create a better world."

—NexGen Systemology Society
Publication Staff
Summer Solstice, June 2020

:: APPENDIX ::
NEXGEN SYSTEMOLOGY GLOSSARY VER. 4.2

—A—

A-for-A (one-to-one) : meaning that what we say, write, represent, think or symbolize is a direct and perfect reflection of the actual aspect or thing—that "A" is for, means and is equivalent to "A" and not "a" or "q" or "!"

aberration : a deviation from, or distortion in, what is true or right or straight.

abreaction : fully reliving traumatic past experiences in order to purge them of their emotional excess.

acid-test : an extreme conclusive process to determine the reality, genuineness or truth of a substance or material. This metaphor refers to a process of applying harsh nitric acid to a golden substance (sample) to determine its genuineness.

acknowledgment : a response-communication establishing that an immediately former communication was properly received, duplicated and understood; the formal acceptance and/or recognition of a communication or presence.

activating event : an incident or occurrence that automatically stimulates a conscious or unrecognized reminder or 'ping' from an earlier *imprinting incident* recorded on one's own personal timeline as an emotionally charged and encoded memory; an incident or instance when thought systems are activated to determine the consequence or significance of an activity, motion or event—often demonstrated as *Activating Event → Belief Systems → Consideration*.

actualization : to make actual; to bring into Reality; to realize fully in *Awareness*.

affinity : the apparent and energetic *relationship* between substances or bodies; the degree of *attraction* or repulsion between things based on natural forces; the *similitude* of frequencies or waveforms; the degree of *interconnection* between systems.

agreement : unanimity of opinion; an accepted arrangement; "reality."

allegorical : a representation of the abstract, metaphysical or "spiritual" using physical or concrete forms.

alpha : the first, primary, basic, superior or beginning of some form; in NexGen system-ology, it also refers to the state of existence that operates on archetypes, will and intention "exterior" to the low-level condensation and solidarity of energy and matter as the 'physical universe'.

alpha control center (ACC) : the highest relay point of *Beingness* for an individuated *Alpha-Spirit, Self* or "I-AM"; in *NexGen Systemology*—a point of spiritual separation of ZU at (7.0) from the *Infinity of Nothingness* (8.0); the truest actualization of *Identity*; the highest *Self-directed* relay of *Alpha-Self* as an *Identity-Continuum*, operating in an *alpha-existence* (or "Spiritual Universe"–AN) to *determine* "Alpha Thought" (6.0) and WILL-*Intention* (5.0) *exterior* to the "Physical Universe"–(KI); the "wave-peak" of "I" emerging as individuated consciousness from *Infinity*.

alpha-spirit : a "spiritual" *Life*-form; the "true" *Self* or I-AM; the spiritual (*alpha*) *Self* that is animating the (*beta*) physical body or "*genetic vehicle*" using a continuous *Lifeline* of spiritual ("*ZU*") energy; an individual spiritual (*alpha*) entity possessing no physical mass or measurable waveform (motion) in the Physical Universe as itself, so it animates the (*beta*) physical body or "*genetic vehicle*" as a catalyst to experience *Self*-determined causality in effect within the *Physical Universe*.

amplitude : the quality of being *ample*; the size or amount of energy that is demonstrated in a *wave*. In the case of audio waves, we associate amplitude with "volume." It is not a statement about the frequencies of waves, only how "loud" they are—to what extent they are or may be projected (or audible).

AN : an ancient cuneiform sign designating the 'spiritual zone'; the *Spiritual Universe*—comprised of spiritual matter and spiritual energy; a direction of motion toward spiritual *Infinity*, away from or superior to the physical ('*KI*'); the spiritual condition of existence providing for our primary *Alpha* state as an individual *Identity* or *I-AM-Self* which interacts and experiences *Awareness* of a *beta* state in the *Physical Universe* ('*KI*') as *Life*.

anathema : a thing or person to be detested, loathed or avoided; a thing or person accursed or despised such as to wish damnation or "divine punishment" upon.

anchor (*conceptual*) : a stable point in space; a fixed point used to hold or stabilize a spatial existence of other points; a spatial point that fixes the parameters of dimensional orientation, such as the corner-points of a solid object in relation to other points in space; in *NexGen Systemology*, "beta-anchored" is an expression used to describe the fixed orientation of a viewpoint from Self in relation to all possible spatial points in *beta-existence* ("physical universe"), or else the existential points that fix the operation of the "body" within the space-time of *beta-existence*.

Ancient Mystery School : the original arcane source of all esoteric knowledge on Earth, concentrated between the Middle East and modern-day Turkey and Transylvania c. 6000 B.C. and then dispersing south (Mesopotamia), west (Europe) and east (Asia) from that location.

antinomian : a term applied to *Gnostics* (popularized by Martin Luther during the Christian reformation) denoting a rejection of formal religious morals and dogma—decreed, written and interpreted by humanity—as a true pathway to Ascension (some elements appear in all forms of religious protest and reformation but as an extreme, would be considered spirto-religious rebellious punkdom by some modern standards, but it should be understood that it does follow a higher ethic, such as Mardukite Utilitarianism.

apotheosis : from the *Greek* word, meaning *"to deify"*; the highest point or apex (for example, of "true knowledge" and "true experience"); an ultimate development of; a glorified or "deified" *ideal*, such as is a quality of *godhood*.

apparent : visibly exposed to sight; evident rather than actual, as presumed by Observation; readily perceived, especially by the senses.

a-priori : from "cause" to "effect"; from a general application to a particular instance; existing in the mind prior to, and independent of experience or observation; validity based on consideration and deduction rather than experience.

archetype : a "first form" or ideal conceptual model of some aspect; the ultimate prototype of a form on which all other conceptions are based.

ascension : actualized *Awareness* elevated to the point of true "spiritual existence" exterior to *beta existence*. An "Ascended Master" is one who has returned to an incarnation on Earth as an inherently *Enlightened One*, demonstrable in their actions—they have the ability to *Self-direct* the "Spirit" as *Self*, just as we are treating the "Mind" and "Body" at this current grade of instruction; in *Moroii ad Vitam*, a state of Beingness after *First Death*, experienced by an *etheric body*, which is able to maintain consciousness as a personal identity continuum with the same *Self-directed* control and communication of Will-Intention that is exercised, actualized and developed deliberately during one's present incarnation.

assessment scale : an official assignment of graded/gradient numeric values.

associative knowledge : significance or meaning of a facet or aspect assigned to (or considered to have) a direct relationship with another facet; to connect or relate ideas or facets of existence with one another; a reactive-response image, emotion or conception that is suggested by (or directly accompanies) something other than itself; in traditional systems logic, an equivalency of significance or meaning between facets or sets that are grouped together, such as in *(a + b) + c = a + (b + c)*; in NexGen Systemology, erroneous associative knowledge is assignment of the same value to all facets or parts considered as related (even when they are not actually so), such as in *a = a, b = a, c = a* and so forth without distinction.

assumption : the act of taking or gather to one's Self; taking possession of.

attenergy : *NexGen Systemological NewSpeak* for "attention energies"; the flow of consciousness energy that is directed as "attention"; a recognition of the axiom from the *Arcane Tablets* that: "energy flows where attention goes."

attention : active use of *Awareness* toward a specific aspect or thing; the act of "attending" with the presence of *Self*; a direction of focus or concentration of *Awareness* along a particular channel or conduit or toward a particular terminal node or communication termination point; the Self-directed concentration of personal energy as a combination of observation, thought-waves and consideration.

authoritarian : knowledge as truth, boundaries and freedoms dictated to an individual by a perceived, regulated or enforced "authority."

auto-suggestion (self-hypnosis) : auto-conditioning; self-programming; delivering directed affirmations or statements repeatedly to *Self* in order to condition a change in behavior or beliefs; any *Self-directed* technique intended to generate a specific "*post-hypnotic suggestion.*"

awareness : the highest sense of-and-as Self in knowing and being as I-AM (the *Alpha-Spirit*); the extent of beingness directed as a POV experienced by Self as knowingness.

—B—

Babylonian : the Mesopotamian civilization that evolved from *Sumer*; the inception of all societal and religious systematization.

band : a division or group; in *NexGen Systemology*—a division or set of frequencies on the ZU-line that are tuned closely together and referred to as a group.

BAT (Beta-Awareness Test) : a method of *psychometric evaluation* developed for *Mardukite Systemology* to determine a "basic" or "average" state of personal *beta-Awareness*.

"bell, book & candle" : three dissimilar objects that are kept accessible during a processing session (the book is often a copy of *The Systemology Handbook* or a hardcover copy of *The Tablets of Destiny* with the dust-jacket removed if it is less distracting that way); a term meant to indicate a Pilot's "objective processing kit" of objects generally present in the session room (accessible on a shelf, table or pedestal stands); in *NexGen Systemology*, the name of an objective processing philosophy pertaining to command of personal reality; historically, a formal ritual used by the Roman Catholic church to ceremonially declare an individual "guilty of the most heinous sins" as "excommunicated (to hold no further communications with) by anathema"—whereby a *bell* is rung, a *holy book* is closed and all *candles* are snuffed out—thus we therapeutically use the same symbolism historically representing religious fragmentation for modern systematic defragmentation purposes.

beta (awareness) : all consciousness activity ("*Awareness*") in the "Physical Universe" (KI) or else *beta-existence*; *Awareness* within the range of the *genetic-body*, including material thoughts, emotional responses and physical motors; personal *Awareness* of physical energy and physical matter moving through physical space and experienced as "time"; the *Awareness* held by *Self* that is restricted to a physical organic *Lifeform* or "genetic vehicle" in which it experiences causality in the *Physical Universe*.

beta (existence) : all manifestation in the "Physical Universe" (KI); the "Physical" state of existence consisting of vibrations of physical energy and physical matter moving through physical space and experienced as "time"; the conditions of *Awareness* for the *Alpha-spirit* (*Self*) as a physical organic *Lifeform* or "genetic vehicle" in which it experiences causality in the *Physical Universe*.

biological unconsciousness : the organism independent of the sentient *Awareness* of the *Self* to direct it; states induced by severe injury and anesthesia.

biomagnetic/biofeedback : a measurable effect, such as a change in electrical resistance, that is produced by thoughts, emotions and physical behaviors which generate specific 'neurotransmitters' and biochemical reactions in the brain, body and across the skin surface.

—C—

cacophony : dissonant, turbulent, harsh and/or discordant sound or noise.

capable : the actual capacity for potential ability.

catalog : a systematic list of knowledge or record of data.

catalyst : something that causes action between two systems or aspects, but which itself is unaffected as a variable of this energy communication; a medium or intermediary channel.

catharsis / cathartic processing : from the Greek root meaning "pure" or "perfect"; the emptying out or discharge of emotional stores; practices of "consolamentum" where an individual removes distorting/fragmented emotional charges and encoding from a personal energy flow/circuit and some other terminal, thing, &tc.

causative : as being the cause.

chakra : (an archaic term used by ancient wisdom traditions); an etheric wheel-mechanism that processes *ZU* energy at specific frequencies along the *ZU-line*, of which a Human being reportedly has *seven* at various degrees.

channel : a specific stream, course, direction or route.

charge : to fill or furnish with a quality; to supply with energy; to lay a command upon; in *NexGen Systemology*—to imbue with intention; to overspread with emotion; application of *Self-directed (WILL)* "intention" toward an emotional manifestation in beta-existence; personal energy stores and significances entwined as fragmentation in mental images, reactive-response encoding and intellectual (and/or) programmed beliefs; in traditional mysticism, to intentionally fix an energetic resonance to meet some degree, or to bring a specific concentration of energy that is transferred to a focal point, such as an object or space.

circuit : see *"feedback loop."*

chronologically : concerning or pertaining to "time."

clockwork : rigidly fixed gear-like systems that operate mechanically and directly upon one another to function.

codification : the process of arranging knowledge in a systematic form.

collapsing a wave : also, *"wave-function collapse"*; the definition or calculation of a wave-function or interaction of potential interactions by Observation. The idea that the Observer is collapsing the wave-function by measuring it appears in the field of *Quantum Physics*. Consciousness or *Awareness* "collapses" the wave-function of energy and matter as the third (required) Principle of Apparent Manifestation; potentiality as a wave is collapsed into an apparent *"is"* (which may be undone by "flattening" the wave back into its state of potentiality).

command : in *Metahuman Systemology*, abilities of the Self (I-AM), from its ideal exterior POV as Alpha Spirit, to direct a communication for control that is perfectly duplicated along the ZU-line without fragmentation.

command line : see "*processing command line.*"

common knowledge (game theory) : facts that all "players" know, and they know that all other "players" also know—such as the very structure of the "game" being played.

communication : successful transmission of information, data, energy (&tc.) along a message line, with a reception of feedback; an energetic flow of intention to cause an effect (or duplication) at a distance; the personal energy moved or acted upon by will or else 'selective directed attention'; the 'messenger action' used to transmit and receive energy across a medium.

compulsion : a failure to be responsible for the dynamics of control—starting, stopping or altering—on a particular channel of communication and/or regarding a particular terminal in existence; an energetic flow with the appearance of being 'stuck' on the action it is already doing or by the control of some automatic mechanism.

computing device : a calculator or modern computer.

condense (condensation) : the transition of vapor to liquid; a change in state demoting a more substantial or solid condition; leading to a more compact or solid form.

condition : an apparent state; circumstances and dynamics that affect the order and function of a system; a series of interconnected requirements, barriers and allowances that must be met; in traditional language, to bring a thing toward a specific, desired or intentional new state (such as in conditioning), though to minimize confusion, *NexGen Systemology* usually treats this with the semantics of imprinting, encoding and programming.

conflict : the opposition of two forces of similar magnitude along the same channel or competing for the same terminal; the inability to duplicate another POV; a thought, intention or communication that is met with an opposing counter-thought or counter-intention that generates an energetic cluster.

confront : to come around in front of; to be in the presence of; to stand in front of, or in the face of; to meet "face-to-face" or "face-up-to."

consciousness : the energetic flow of *Awareness*; the Principle System of *Awareness* that is spiritual in nature, which demonstrates potential interaction with all degrees of the

Physical Universe; the *Beingness* component of our existence in *Spirit*; the Principle System of *Awareness* as *Spirit* that directs action in the Mind-System.

consensual : formed or existing simply by consent—by general or mutual agreement; permitted, approved or agreed upon by majority of opinion; knowingly agreed upon unanimously by all concerned; to be in agreement on the objective universe and/or a course of action therein.

consideration : careful analytical reflection of all aspects; deliberation; determining the significance of a "thing" in relation to similarity or dissimilarity to other "things"; evaluation of facts and importance of certain facts; thorough examination of all aspects related to, or important for, making a decision; the analysis of consequences and estimation of significance when making decisions.

continuity : being a continuous whole; a complete whole or "total round of"; the balance of the equation [$-120 + 120 = 0$, &tc.]

continuum : a continuous *whole*; observing all gradients on a *spectrum*; measuring quantitative variation with gradual transition on a spectrum without demonstrating discontinuity or separate parts.

control (systems) : Communication relayed from an operative center or organizational cluster, which incites new activity elsewhere in a system.

correlating : the relationship between two or more aspects, parts or systems.

correspondence : a direct relationship or correlation; see also *"associative knowledge."*

Cosmic History : the entire timeline of all existence, starting with the Infinity of Nothingness and individuation of Self and ending with the condensation and solidification of this Physical Universe experienced in present-time.

Cosmic Law : the "Law" of Nature (or the Physical Universe); the "Law" governing cosmic ordering; often called "Natural Law" in sciences and philosophies that attempt to codify or systematize it.

cosmology : a philosophy defining the origins and structure of the universe.

Cosmos : archaic term for the "physical universe"; implies that chaos was brought into order; includes past "universes" that we occupied.

counter-productive : contrary to the greater purpose; anything which brings *all Life* away from its sustainable goals of *Infinite Existence.*

crash-coursed : a very intense or steep delivery of education over a very brief time period.

Crossing the Abyss : to enter the spiritual or metaphysical unknown in "Self-annihilation" to purify the Self and "return to the Source."

cuneiform : the oldest extant writing from Mesopotamia; wedge-shaped script inscribed on clay tablets with a reed pen.

cuneiform signs : the cuneiform script, as used in ancient Mesopotamia, is not represented in a linear alphabet of "letters," but by a systematic use of basic word "signs" that are combined to form more complex word "signs."

—D—

data-set : the total accumulation of knowledge used to base Reality.

dead-memories : outdated, inadequate or erroneous data.

defragmentation : the *reparation* of wholeness; a process of removing *"fragmentation"* in data or knowledge to provide a clear understanding; applying techniques and processes that promote a *holistic* interconnected *alpha* state, favoring observational *Awareness* of continuity in all spiritual and physical systems; in *NexGen Systemology*, a *"Seeker"* achieving an actualized state of basic *"Self-Honest Awareness"* is said to be *defragmented.*

degree : a physical or conceptual *unit* (or point) defining the variation present relative to a *scale* above and below it; any stage or extent to which something *is* in relation to other possible positions within a *set* of *"parameters"*; a point within a specific range or spectrum; in *NexGen Systemology*, a *Seeker's* potential energy variations or fluctuations in thought, emotional reaction and physical perception are all treated as *"degrees."*

demographics : identifying segments of the population, real or representative.

destiny : what is set down, made firm, standard, or stands fixed as a constant end; the absolute *destination* regardless of whatever course is traveled; in *NexGen Systemology*, the *"destiny"* of the *"Human Spirit"* (or *"Alpha Spirit"*) is infinite existence—*"Immortality."*

dichotomy : a division into two parts, types or kinds.

differential : the quantitative value difference between two forces, motions, pressures or degrees.

differentiation : an apparent difference between aspects or concepts.

discernment : to perceive, distinguish and/or differentiate experience into true knowledge.

displace : to compel to leave; to move or replace something with something else in its place or space.

dissonance : discordance; out of step; out of phase; disharmonious; the "differential" between the way things are and the way things are experienced; cognitive dissonance could be demonstrated as A = abc, or C = A, the duplication of truth/communication is not A-for-A.

dross : prime material; specifically waste-matter or refuse; the discarded remains collected together.

dynamic (systems) : a principle or fixed system which demonstrates its *'variations'* in activity (or output) only in constant relation to variables or fluctuation of interrelated systems; a standard principle, function, process or system that exhibits *'variations'* and change simultaneously with all connected systems.

—E—

Eastern traditions : the evolution of the *Ancient Mystery School* east of its origins, primarily the Asian continent, or what is archaically referred to as "oriental."

echelon : a level or rung on a ladder; a rank or level of command.

eclipse : to cast a shadow or darken; to block out or obscure a comparison.

elocution : the skillful use of clearly directed and expressive speech; the expert demonstration of articulation, pronunciation and dictation to express a message.

emotional encoding : the substance of *imprints*; associations of sensory experience with an *imprint*; perceptions of our environment that receive an *emotional charge*, which form or reinforce facets of an *imprint*; perceptions recorded and stored as an *imprint* within the "emotional range" of energetic manifestation; the formation of an energetic store or charge on a channel that fixes emotional responses as a mechanistic automation, which is carried on in an individual's spiritual timeline or personal continuum of existence.

enact : to make happen; to bring into action; to make part of an act.

encompassing : to form a circle around, surround or envelop around.

energetic exchange : communicated transmission of energetically encoded "information" between fields, forces or source-points that share some degree of interconnectivity; the event of "waves" acting upon each other like a force, flowing in regard to their proximity, range, frequency and amplitude.

energy signatures : a distinctive pattern of energetic action.

enforcement : the act of compelling or putting (effort) into force; to compel or impose obedience by force; to impress strongly with applications of stress to demand agreement or validation; the lowest-level of direct control by physical effort or threat of punishment; a low-level method of control in the absence of true communication.

engineering : the *Self-directed* actions and efforts to utilize knowledge (observed causality/science), maths (calculations/quantification) and logic (axioms/formulas) to understand, design or manifest a solid structure, machine, mechanism, engine or system; as "*Reality Engineering*" in *NexGen Systemology*—intentional *Self-directed* adjustment of existing Reality conditions; the application of total *Self-determinism* in *Self-Honesty* to change apparent Reality using fundamentals of *Systemology* and *Cosmic Law*.

entanglement : tangled together; intertwined and enmeshed systems; in *NexGen Systemology*, a reference to the interrelation of all particles as waves at a higher point of connectivity than is apparent, since wave-functions only "collapse" when someone is *Observing*, or doing the measuring, evaluating, &tc.

entropy : the reduction of organized physical systems back into chaos-continuity when their integrity is measured against space over time.

epicenter : the point from which shock-waves travel.

epistemology : a school of philosophy focused on the truth of knowledge *and* knowledge of truth; theories regarding validity and truth inherent in any structure of knowledge and reason.

erroneous : inaccurate; incorrect; containing error.

esoteric : hidden; secret; knowledge understood by a select few.

etching : to cut, bite or corrode with acid to produce a pattern.

evaluate : to determine, assign or fix a set value, amount or meaning.

exacting : a demanding rigid effort to draw forth from.

executable : the supreme authoritative ability to carry out according to design.

existence : the *state* or fact of *apparent manifestation*; the resulting combination of the Principles of Manifestation: consciousness, motion and substance; continued *survival*; that which independently exists; the 'Prime Directive' and sole purpose of all manifestation or Reality; the highest common intended motivation driving any "*Thing*" or *Life*.

existential : pertaining to existence, or some aspect or condition of existence.

extant : in existence; existing.

exoteric : public knowledge or common understanding; the level of understanding and *Knowing* maintained by the "masses"; the opposite of *esoteric*.

experiential data : accumulated reference points we store as memory concerning our "experience" with Reality.

extrapolate : to make an estimate of the "value" outside of the perceivable range.

extropy : in *NexGen Systemology*—the reduction of organized spiritual systems back into a singularity of Infinity when their integrity is measured against space over time.

—F—

facets : an aspect, an apparent phase; one of many faces of something; a cut surface on a gem or crystal; in *NexGen Systemology*—a single perception or aspect of a memory or "Imprint"; any one of many ways in which a memory is recorded; perceptions associated with a painful emotional (sensation) experience and "*imprinted*" onto a metaphoric lens through which to view future similar experiences; other secondary terminals that are associated with a particular terminal, painful event or experience of loss, and which may exhibit the same encoded significance as the activating event.

faculties : abilities of the mind (individual) inherent or developed.

fallacy : a deceptive, misleading, erroneous and/or false beliefs; unsound logic; persuasions, invalidation or enforcement of Reality agreements based on authority, sympathy, bandwagon/mob mentality, vanity, ambiguity, suppression of information, and/or presentation of false dichotomies.

fate : what is brought to light or actualized as experience; the actual *course* taken to reach an end, charted end, or final *destination*; in *NexGen Systemology*, the *'fate'* of a 'Human Spirit' (or 'Alpha Spirit') is determined by the choice of course taken to experience *Life*.

feedback loop : a complete and continuous circuit flow of energy or information directed as an output from a source to a target which is altered and return back to the source as an input; in *General Systemology*—the continuous process where outputs of a system are routed back as inputs to complete a circuit or loop, which may be closed or connected to other systems/circuits; in *NexGen Systemology*—the continuous process where directed *Life* energy and *Awareness* is sent back to *Self* as experience, understanding and memory to complete an energetic circuit as a loop.

flattening a wave : see *"process-out"* for definition; also see *"collapsing a wave."*

flow : movement across (or through) a channel (or conduit); a direction of active energetic motion typically distinguished as either an *in-flow*, *out-flow* or *cross-flow*.

fractal : a wave-curve, geometric figure, form or pattern, with each part representative of the same characteristics as the whole; any baseline, sequence or pattern where the 'whole' is found in the 'parts' and the 'parts' contain the 'whole'; a pattern that reoccurs similarly at various scales/levels on a continuous whole; a subset of a Euclidean space explored in higher-level academic mathematics, in which fractal dimensions are found to exceed topological ones; in NexGen Systemology, a "fractal-like" description is used specifically for a pattern or form that has a reoccurring nature without regard to what level or scale it is manifest upon. Examples include the formation of crystals, tree-like patterns, the comparison of atoms to solar systems to galaxies, &tc.

fragmentation : breaking into parts and scattering the pieces; the *fractioning* of wholeness or the *fracture* of a holistic interconnected *alpha* state, favoring observational *Awareness* of perceived connectivity between parts; *discontinuity*; separation of a totality into parts; in *NexGen Systemology*, a person outside a state of *Self-Honesty* is said to be *fragmented*.

—G—

game : a strategic situation where the power of choice is employed or affected; a parameter or condition defined by purposes, freedoms and barriers (rules).

game theory : a mathematical theory of logic pertaining to strategies of maximizing gains and minimizing loses within prescribed boundaries and freedoms; a field of knowledge widely applied to human problem solving and decision-making; the application of true knowledge and logic to deduce the correct course of action given all variables and interplay of dynamic systems; logical study of decision making where "players" make choices that affect (the interests) of other "players"; an intellectual study of conflict and cooperation.

general systemology ("systematology") : a methodology of analysis and evaluation regarding the systems—their design and function; organizing systems of interrelated information-processing in order to perform a given function or pattern of functions.

genetic memory : the evolutionary, cellular and genetic (DNA) "memory" encoded into a *genetic vehicle* or *living organism* during its progression and duplication (reproduction) over millions (or billions) of years on Earth; in *NexGen Systemology*—the past-life Earth-memory carried in the genetic makeup of an organism (*genetic vehicle*) that is *independent of any* actual "spiritual memory" maintained by the *Alpha Spirit* themselves, from its own previous lifetimes on Earth and elsewhere using other *genetic vehicles* with no direct evolutionary connection to the current physical form in use.

genetic-vehicle : a physical *Life*-form; the physical (*beta*) body that is animated/controlled by the (*Alpha*) *Spirit* using a continuous *Lifeline* (ZU); a physical (*beta*) organic receptacle and catalyst for the (*Alpha*) *Self* to operate "causes" and experience "effects" within the *Physical Universe*.

gifted : attributing a special quality or ability; having exceptionally high intelligence or mental faculties.

gnosis : a *Greek* word meaning knowledge, but specifically "true knowledge"; the highest echelon of "true knowledge" accessible (or attained) only by mystical or spiritual faculties whereby actualized realizations are achieved independent of specialized education.

Gnostics : a name meaning "having knowledge" in Greek language (see also *gnosis*); an early sect of Judeo-Christian mysticism from the 1st Century AD emphasizing true knowledge by *Self-Honest* experience of metahuman and spiritual states of beingness, emphasizing defragmentation of "illusion" and the overcoming of material "deception";

an esoteric proto-Systemology organization disbanded by the Roman Church as heretical.

godhood : a divine character or condition; "divinity."

gradient : a degree of partitioned ascent or descent along some scale, elevation or incline; "higher" and "lower" values in relation to one another.

—H—

heralded : proclaimed ahead of or prior to; officially announced.

holistic : the examination of interconnected systems as encompassing something greater than the *sum* of their "parts."

Homo Novus : literally, the "new man"; the "newly elevated man" or "known man" in ancient Rome; the man who "knows (only) through himself"; in NexGen Systemology—the next spiritual and intellectual evolution of *homo sapiens* (the "modern Human Condition"), which is signified by a demonstration of higher faculties of *Self-Actualization* and clear *Awareness*.

Homo Sapiens Sapiens : the present standard-issue Human Condition; the *hominid* species and genetic-line on Earth that received modification, programming and conditioning by the *Anunnaki* race of *Alpha-Spirits*, of which early alterations contributed to various upgrades (changes) to the genetic-line, beginning approximately 450,000 years ago (*ya*) when the *Anunnaki* first appear on Earth; a species for the Human Condition on Earth that resulted from many specific *Anunnaki* "genetic" and "cultural" *interventions* at certain points of significant advancement—specifically (but not limited to) *circa* 300,000 *ya*, 200,000 *ya*, 40,000 *ya,* and 8,000 *ya*; a species of the Human Condition set for replacement by *Homo Novus*.

Human Condition : a standard default state of Human experience that is generally accepted to be the extent of its potential identity (*beingness*)—currently treated as *Homo Sapiens Sapiens,* but which is scheduled for replacement by *Homo Novus*.

humanistic psychology : a field of academic psychology approaching a holistic emphasis on *Self-Actualization* as an individual's most basic motivation; early key figures from the 20th century include: Carl Rogers, Abraham Maslow, L. Ron Hubbard, William Walker Atkinson, Deepak Chopra and Timothy Leary (to name a few).

hypothetical : operating under the assumption a certain aspect actual "is."

—I—

identity : the collection of energy and matter—including memory—across the "*Spiritual Continuum*" that we consider as "I" of *Self*.

identity-system : the application of the *ZU-line* as "I"—the continuous expression of *Self* as *Awareness*.

illuminated : to supply with light so as to make visible or comprehensible.

imagination : the ability to create imagery in the mind at will and change or alter it as desired; the ability to create, change and dissolve mental images on command or as an act of will; to create a mental image or have associated imagery displayed (or "conjured") in the mind that may or may not be treated as real (or memory recall) and may or may not accurately duplicate objective reality.

immersion : plunged or sunk into; wholly surrounded by.

imprint : to strongly impress, stamp, mark (or outline) onto a softer 'impressible' substance; to mark with pressure onto a surface; in *NexGen Systemology*, the term is used to indicate permanent Reality impressions marked by frequencies, energies or interactions experienced during periods of emotional distress, pain, unconsciousness, loss, enforcement, or something antagonistic to physical (personal) survival, all of which are are stored with other reactive response-mechanisms at lower-levels of *Awareness* as opposed to the active memory database and proactive processing center of the Mind; an experiential "memory-set" that may later resurface—be triggered or stimulated artificially—as Reality, of which similar responses will be engaged automatically.

imprinting incident : the first or original event instance communicated and *emotionally encoded* onto an individual's "timeline" (their memory of events over the course of one or all lifetimes) that formed a permanent impression that is later used to mechanistically treat future contact on that channel; the first or original occurrence of some particular *facet* or mental image related to a certain type of *encoded response*, such as pain and discomfort, losses and victimization, and even the acts that we have taken against others along the timeline that caused them to also be *Imprinted*.

incarnation : a present, living or concrete form of some thing or idea.

inception : the beginning, start, origin or outset.

incite : to urge on or cause; instigate; prove or stimulate into action.

indefinable : without a clear definition being currently presented.

individual : a person, lifeform or human entity; a *Seeker* or potential *Seeker* is often referred to as an "individual" within Mardukite Zuism and Systemology materials.

infinite existence : "immortality."

inhibited : withheld, discouraged or repressed from some state.

"in phase" : see *"phase alignment."*

insistence : repeated use of a communicated energy into a form that demands acknowledgment, is more difficult to avoid or ignore.

institution : a social standard or organizational group responsible for promoting some system or aspect in society.

intention : the directed application of Will; to intend (have "in Mind") or signify (give "significance" to) for or toward a particular purpose; in *NexGen Systemology* (from the *Standard Model*)—the spiritual activity at WILL (5.0) directed by an *Alpha Spirit* (7.0); the application of WILL as "Cause" from a higher order of Alpha Thought and consideration (6.0), which then may continue to relay communications as an "effect" in the universe.

interdimensional : systems that are interconnected or correlated between the Physical Universe and the Spiritual Universe—or between "dimension states" observably identified as "physical," "emotional," "psychological" and "spiritual." The only point of true interconnectivity that we can systematically determine is called *"Life."*

intermediate : a distinct point between two points; actions between two points.

invalidate : decrease the level or degree or *agreement* as Reality.

invests : spends on; gives or devotes something to earn a result; endows with.

—K—

"kNow" : a creative spelling and use of semantics for "know" and "now" to indicate the state of present-time actualized "Awareness" as Self (Alpha-Spirit), developed for fun dual-meaning messages made by early Mardukite Systemologists in 2008-9, such as "live in the kNow" or "be in the kNow" and even "drown in the kNow" &tc.

knowledge : clear personal processing of informed understanding; information (data) that is actualized as effectively workable understanding; a demonstrable understanding on which we may 'set' our *Awareness*—or literally a "know-ledge."

KI : an ancient cuneiform sign designating the *'physical zone'*; the *Physical Universe*—comprised of physical matter and physical energy in action across space and observed as time; a direction of motion toward material *Continuity*, away from or subordinate to the Spiritual ('AN'); the physical condition of existence providing for our *beta* state of *Awareness* experienced (and interacted with) as an individual *Lifeform* from our primary Alpha state of Identity or *I-AM-Self* in the *Spiritual Universe* ('AN').

kinetic : pertaining to the energy of physical motion and movement.

—L—

learned : highly educated; possessing significant knowledge.

level : a physical or conceptual *tier* (or plane) relative to a *scale* above and below it; a significant *gradient* observable as a *foundation* (or surface) built upon and subsequent to other levels of a totality or whole; a *set* of "*parameters*" with respect to other such *sets* along a *continuum*; in *NexGen Systemology*, a *Seeker's* understanding, *Awareness* as *Self* and the formal grades of material/instruction are all treated as "*levels*."

Liber-One : First published in October 2019 as "*The Tablets of Destiny: Using Ancient Wisdom to Unlock Human Potential*" by Joshua Free; republished in the complete *Grade-III* anthology, "*The Systemology Handbook.*"

Liber-2B : First published in December 2019 as "*Crystal Clear: The Self-Actualization Manual & Guide to Total Awareness*" by Joshua Free; republished in the complete *Grade-III* anthology, "*The Systemology Handbook.*"

Liber-2C : First published in April 2020 as "*Communication and Control of Energy & Power: The Magic of Will & Intention (Volume One)*" by Joshua Free; an integral part of the *Grade-IV* Professional Piloting Course.

Liber-2D : First published in June 2020 as "*Command of the Mind-Body Connection: The Magic of Will & Intention*" (Volume Two)" by Joshua Free; an integral part of the *Grade-IV* Professional Piloting Course.

Liber-3C : First published in July 2020 as "*Now You Know: The Truth About Universes & How*

You Got Stuck in One" by Joshua Free; a manual in the *Grade-IV* Metahuman Systemology series.

Liber-3D : A forthcoming manual in the *Grade-IV* Metahuman Systemology series.

localized : brought together and confined to a particular place.

logic equations : using symbols and basic mathematical logic to establish the validity of statements or to see how a variable within a system will change the result; a basic demonstration of proportion or relationship between variables in a system.

logistics : pertaining to the movement or transportation between locations.

—M—

macrocosmic : taking examples and system demonstrations at one level and applying them as a larger demonstration of a relatively higher level or unseen dimension.

manifestation : something brought into existence.

Marduk : founder of Babylonia; patron Anunnaki "god" of Babylon.

Mardukite Zuism : a Mesopotamian-themed (Babylonian-oriented) religious philosophy and tradition applying the spiritual technology based on *Arcane Tablets* in combination with "Tech" from *NexGen Systemology*; first developed in the New Age underground by Joshua Free in 2008 and realized publicly in 2009 with the formal establishment of the "*Mardukite Chamberlains.*"

master control center (MCC) : a perfect computing device to the extent of the information received from "lower levels" of sensory experience/perception; the proactive communication system of the "*Mind*"; a relay point of active *Awareness* along the Identity's *ZU-line*, which is responsible for maintaining basic *Self-Honest Clarity* of *Knowingness* as a *seat of consciousness* between the *Alpha-Spirit* and the secondary "*Reactive Control Center*" of a *Lifeform* in *beta existence*; the Mind-center for an *Alpha-Spirit* to actualize cause in the *beta existence*; the analytical *Self-Determined* Mind-center of an *Alpha-Spirit used* to project *Will* toward the genetic body; the point of contact between *Spiritual Systems* and the *beta existence*; presumably the "*Third Eye*" of a being connected directly to the *I-AM-Self*, which is responsible for *determining* Reality at any time; in *NexGen Systemology*, this is plotted at (4.0) on the continuity model of the *ZU-line*.

"Master Grades" : literary materials by Joshua Free (written between 1995 and 2019) revised and compiled for the International School of Systemology instructional grades —"Route of Magick & Mysticism" (*Grade I, Part A*), "Route of Druidism & Dragon Legacy" (*Grade I, Part D*), "Route of Mesopotamian Mysteries" (*Grade II*) and "Route of Mardukite Systemology" (*Grade III*), collectively known as the "Pathway to Self-Honesty."

Mesopotamia : land between Tigris and Euphrates River; modern-day Iraq.

methodology : a system of methods, principles and rules to compose a systematic paradigm of philosophy or science.

"Mind's Eye" : the point where the "mental pictures" (and senses) are generated that define what an individual believes they are experiencing in present time; the activities of the "Third-Eye" (or actualized MCC) where the *Alpha-Spirit* directly interacts with the organic *genetic vehicle* in *beta-existence*; *Self-directed* activity on the plane of "mental consciousness" maintained between "spiritual consciousness" of the *Alpha-Spirit* and the "physical/emotional consciousness" of the *genetic vehicle*; the "consciousness activity" *Self-directed* by an actualized WILL.

misappropriated : put into use incorrectly; to apply ineffectively or as unintended by design.

motor functions : internal mechanisms that allow a body to move.

—N—

Nabu : the original "god of wisdom, writing and knowledge." (Babylonian)

negligible : so small or trifle that it may be disregarded.

neurotransmitter : a chemical substance released at a physiological level (of the genetic vehicle) that bridges communication of energetic transmission between the *Mind-Body* systems, using the "nervous system" of the physical body; biochemical amino acids and peptides (neuropeptides), hormones, &tc.

NexGen Systemology : a modern tradition of applied religious philosophy and spiritual technology based on *Arcane Tablets* in combination with "*general systemology*" and "*games theory*" developed in the New Age underground by Joshua Free in 2011 as an advanced futurist extension of the "*Mardukite Chamberlains.*"

—O—

objectively : concerning the "external world" and attempts to observe Reality independent of personal "subjective" factors.

optimum : the most favorable or ideal conditions for the best result; the greatest degree of result under specific conditions.

orchestration : to arrange or compose the performance of a system.

"Ordo Templi Orientis" (OTO) : otherwise "Order of the Eastern Temple" or "Order of the Eastern Templars"; a underground occult German-inspired variation on masonic-styled "Golden Dawn and Rosicrucian type" methodology; founded by Theodor Reuss (*et al.*) in *c.*1900, and within a decade included such membership as Gerald Gardner, Aleister Crowley and other early prominent "New Age" figures; surviving mainstream OTO factions are those strongly influenced by Crowley's "Thelemic" work and later "Typhonian" contributions by his heir-apparent, Kenneth Grant.

organic : as related to a physically living organism or carbon-based life form; energy-matter condensed into form as a focus or POV of Spiritual Life Energy (*ZU*) as it pertains to beta-existence of *this* Physical Universe (*KI*).

oscillation-alternation : a particular type of (or fluctuation) between two relative states, conditions or degrees; a wave-action between two degrees, such as is described in the action of the *pendulum effect*; a flux or wave-like energy in motion, across space, calculable as time; in systematic processing, alternation is the shift between two direction flows on a circuit channel, such as *inflow* and *outflow*, or between two types of processing, such as *objective* and *subjective*; alternation of a POV creates "space."

—P—

pantheism : religious philosophies that observe God as inherent within all aspects of the Physical Universe.

paradigm : an all-encompassing *standard* by which to view the world and *communicate* Reality; a standard model of reality-systems used by the Mind to filter, organize and interpret experience of Reality.

parameters : a defined range of possible variables within a model, spectrum or continuum; the extent of communicable reach capable within a system or across a distance; the defined or imposed limitations placed on a system or the functions within a system;

the extent to which a Life or "thing" can *be*, *do* or *know* along any channel within the confines of a specific system or spectrum of existence.

paramount : the most important; "above all else."

participation : being part of the action or affecting the result.

patterns (probability patterns) : observation of cycles and tendencies to predict a causal relationship or determine the actual condition or flow of dynamic energy using a holistic systemology to understand Life, Reality and Existence as opposed to isolating or excluding perceived parts as being mutually separate from other perceived parts.

patron god : the most sacred deity of a region or city, of which most temples and religious services are directed; the personal deity of an individual.

PCL : see "*processing command line.*"

personality (program) : the total composite picture an individual "identifies" themselves with; the accumulated sum of material and mental mass by which an individual experiences as their timeline; a "beta-personality" is mainly attached to the identity of a particular physical body and the total sum of its own genetic memory in combination with the data stores and pictures maintained by the Alpha Spirit; a "true personality" is the Alpha Spirit as Self completely defragmented of all erroneous limitations and barriers to consideration, belief, manifestation and intention.

perturbation : the deviation from a natural state, fixed motion, or orbit system caused by another external system; disturbing or disquieting the serenity of an existent state; inciting observable apparent action using indirect or outside actions or 'forces'; the introduction of a new element or facet that disturbs equilibrium of a standard system; the "butterfly effect"; in *NexGen Systemology*, 'perturbation' is a necessary condition for the *ZU-line* to function as a *Standard Model* of actual *'monistic continuity'*—which is a *Lifeforce* singularity expressed along a spectrum with potential interactions at each degree from any source; the influence of a degree in one state by activities of another state that seem independent, but which are actually connected directly at some higher degree, even if not apparently observed.

phase : in *NexGen Systemology,* a pattern of personality or identity that is assumed as the POV from *Self;* see also "*phase alignment.*"

phase alignment or "***in phase***" : to be in synch, in step or aligned properly with something else in order to increase the total strength value; in *NexGen Systemology*—referring to alignment of *Awareness* with a particular identity, space or time (such as being *in* Self *in* present *space* and *time*).

philanthropy : charitable; the intention (or programmed desire) to generously provide personal wealth and service to the well-being and continued existence of others.

physics : a science of motions, forces and bodies in the Physical Universe.

physiology : a science of motions of living bodies or organisms.

pilfering : to steal in small quantities; petty theft.

pilot : the steersman of a ship; in *NexGen Systemology*—an individual qualified to operate *Systemology Processing* for other *Seekers* on the *Pathway to Self-Honesty*.

ping : a short, high pitched ring, chime or noise that alerts to the presence of something; in computer systems, a query sent on a network or line to another terminal in order to determine if there is a connection to it; in *NexGen Systemology*, the sudden somatic twinge or pain or discomfort that is felt as a sensation in the body when a particular terminal (lifeform, object, concept) is 'brought to mind' or contacted on a personal communication channel-circuit; the accompanying sensations and mental images that are experienced as an automatic-response to the presence of some channel or terminal.

player (game theory) : an individual that is making decisions in a game and/or is affected by decisions others are making in the game, especially if those other-determined decisions now affect the possible choices.

point-of-view (POV) : an opinion or attitude as expressed from a specific identity-phase; a specific standpoint or vantage-point; a definitive manner of consideration specific to an individual phase or identity; a place or position affording a specific view or vantage; circumstances and programming of an individual that is conducive to a particular response, consideration or belief-set (paradigm); a position (consideration) or place (location) that provides a specific view or perspective (subjective) on experience (of the objective).

postulate : to put forward as truth; to suggest or assume an existence *to be*; to provide a basis of reasoning and belief; a basic theory accepted as fact.

potentiality : the total "sum" (collective amount) of "latent" (dormant—present but not apparent) capable or possible realizations; used to describe a state or condition of what has not yet manifested, but which can be influenced and predicted based on observed patterns and, if referring to beta-existence, Cosmic Law.

precedent : a matter which precedes or goes before another in importance.

precipitate : to actively hasten or quicken into existence.

preconception : to assign values or evaluate a reaction or response to a past "imprint" of something and treat it as present knowledge or experience.

prehistoric : any time before human history is written; prior to c. 4000 B.C.

premise : a basis or statement of fact from which conclusions are drawn.

presence : personal orientation of Self located in space and time and handling the energy-matter present; the quality of some thing (energy/matter) being "present" in space-time.

prevalent : of wide extent; an extensive or largely accepted aspect or current state.

probability : the causal likelihood for something to result, "effect" or manifest in and as a certain way, manner or degree, based on "observed evaluation" of programming and tendencies that follow Cosmic Law.

"process-out" or **"flatten a wave"** : to reduce *emotional encoding* of an *imprint* to zero; to dissolve a *wave-form* or *thought-formed* "solid" such as a "*belief*"; to completely run a *process* to its end, thereby *flattening* any previously "*collapsed-waves*" or *fragmentation* that is obstructing the *clear channel* of *Self-Awareness*; also referred to as "processing-out"; to discharge all previously held emotionally encoded imprinting or erroneous programming and beliefs that otherwise fix the free flow (wave) to a particular pattern, solid or concrete "*is*" form.

processing, systematic : the inner-workings or "through-put" result of systems; in *Nex-Gen Systemology*, a methodology of applied spiritual technology used toward personal Self-Actualization; methods of selective directed attention, communicated language and associative imagery that targets an increase in personal control of the human condition.

processing command line (PCL) or **command line** : a directed input; a specific command using highly selective language for *Systemology Processing*; a predetermined directive statement (cause) intended to focus concentrated attention (effect).

projecting awareness : sending out (motion) or radiating *"consciousness"* from *Self* ("I") to another POV.

proportional : having a direct relationship or mutual interaction with.

protest : a communication objecting the enforcement or rejection of a prior communication; an effort the cancel the "is" quality, or existence of a previous creation or communication; the unwillingness to be the POV of effect.

Proto-Indo-European (PIE) : a single source root language c.4500 B.C. contributing to most European languages.

psychometric evaluation : the relative measurement of personal ability, mental (psychological/thought) faculties, and effective processing of information and external stimulus data; a scale used in "applied psychology" to evaluate and predict human behavior.

—R—

rationality / reasoning (game theory) : the extent to which a player seeks to play (make decisions, &tc.) in order to maximize the gains (or else survival) achievable within any given game conditions; the ability and willingness of an individual to reach toward conditions that promote the highest level of survival and existence and make the best choices and moves to see the desired goal manifest.

reactive control center (RCC) : the secondary (reactive) communication system of the *"Mind"*; a relay point of *Awareness* along the Identity's *ZU-line*, which is responsible for engaging basic motors, biochemical processes and any *programmed automated responses* of a living *beta* organism; the reactive Mind-Center of a living organism relaying communications of *Awareness* between causal experience of *Physical Systems* and the *"Master Control Center"*; it presumably stores all emotional encoded imprints as fragmentation of "chakra" frequencies of *ZU* (within the range of the *"psychological/emotive systems"* of a being), which it may *react* to as Reality at any time; in *NexGen Systemology*, this is plotted at (2.0) on the continuity model of the *ZU-line*.

realization : the clear perception of an understanding; to make "real" or give "reality" to so as to grant a property of "beingness" or "being as it is"; the state or instance of coming to an *Awareness*; in *NexGen Systemology*, "gnosis" or true knowledge achieved during *systematic processing*; achievement of a new (or "higher") cognition, true knowledge or perception of Self in relation to reality.

receptacle : a device or mechanism designed to contain and store a specific type of aspect or thing; a container meant to receive something.

recursive : repeating by looping back onto itself to form continuity; *ex.* the "Infinity" symbol is recursive.

relative : an apparent point, state or condition treated as distinct from others.

relinquish : to give up control, command or possession of.

repetitively : to repeat "over and over" again; or else "repetition."

responsibility : the *ability* to *respond*; the extent of mobilizing *power* and *understanding* an individual maintains as *Awareness* to enact *change*; the proactive ability to *Self-direct* and make decisions independent of an outside authority.

resurface : to return to, or bring up to, the "surface" what has been submerged; in *NexGen Systemology*—relating specifically to processes where a *Seeker* recalls blocked energy stored covertly as emotional "*imprints*" (by the RCC) so that it may be effectively defragmented from the "*ZU-line*" (by the MCC).

—S—

scions : a descendant or child offspring; an offshoot or branch.

Seeker : an individual on the *Pathway to Self-Honesty*; a practitioner of *Mardukite Systemology* or *NexGen Systemology Processing* that is working toward *Ascension.*

Self-actualization : bringing the full potential of the Human spirit into Reality; expressing full capabilities and creativeness of the *Alpha-Spirit.*

Self-determinism : the freedom to act, clear of external control or influence; the personal control of Will to direct intention.

Self-evaluation : see "*psychometric evaluation.*"

Self-honesty : the *alpha* state of *being* and *knowing*; clear and present total *Awareness* of-and-as *Self*, in its most basic and true proactive expression of itself as *Spirit* or *I-AM*—free of artificial attachments, perceptive filters and other emotionally-reactive or mentally-conditioned programming imposed on the human condition by the systematized physical world.

self-sustained : self-supported.

semantics : the *meaning* carried in *language* as the *truth* of a "thing" represented, *A-for-A*; the *effect* of language on *thought* activity in the Mind and physical behavior; language as *symbols* used to represent a concept, "thing" or "solid."

semantic-set : the implied meaning behind any groupings of words or symbols used to define a specific paradigm.

sentient : consciously intelligent.

simulacrum : an tangible image, facsimile or superficial representation that carries a likeness or similarity to someone or something else; in *NexGen Systemology*—the *genetic vehicle* or physical body is an example of a "simulacrum" of the true *Alpha-Spirit* or *Self* (I-AM), which otherwise has no tangible form in *beta-existence*.

sine-wave : the *frequency* and amplitude of a quantified (calculable) *vibration* represented on a graph (graphically) as smooth repetitive *oscillation* of a *waveform*; a *waveform* graphed for demonstration—otherwise represented in *NexGen Systemology* logic equations as 'Wf,' or in mathematics as the '*function of x*' (*fx*); graphically representing arcs (*parameters*) of a circular *continuity* on a *continuum*; in the *Standard Model of NexGen Systemology*, the actual 'wave vibration' graphically displayed on an otherwise static *ZU-line* (of Infinity) is a '*sine-wave*'.

singularity : a point where apparently dissimilar qualities of all aspects share a singular expression, nature or quality.

slate : a flat surface used for writing on; a chalk-board.

somatic : pertaining to the physical body and its response actions.

space : the viewpoint (or POV) extended out from any point out toward a dimension or dimensions; the consideration of a point or spot; the field of energy/matter mass created as a result of communication and control in action and measured as time (wavelength).

spectrum : a broad range or array as a continuous series or sequence; defined parts along a singular continuum.

standard issue : equally dispensed to all without consideration.

standard model : a fundamental *structure* or symbolic construct used to evaluate a complete *set* in *continuity* relative to itself and variable to all other *dynamic systems* as graphed or calculated by *logic*; in *NexGen Systemology*—our existential and cosmological cabbalistic model; a "*monistic continuity model*" demonstrating *total system* interconnectivity "above" and "below" observation of any apparent *parameters*; the *ZU-line* represented as a singular vertical (*y*-axis) waveform in space across dimensional levels (universes) without charting any specific movement across a dimensional time-graph *x*-axis.

static : characterized by a fixed or stationary condition; having no apparent change, movement or fluctuation.

stoicism : pertaining to the school of "stoic" philosophy, distinguished by calm mental attitudes, freedom from desire/passion and essentially any emotional fluctuation.

sub-zones : at ranges "below" which we are representing or which is readily observable for current purposes.

successively : what comes after; forward into the future.

succumb : to give way, or give in to, a relatively stronger superior force.

Sumerian : ancient civilization of *Sumer*, founded in Mesopotamia c. 5000 B.C.

superfluous : excessive; unnecessary; needless.

superstition : knowledge accepted without good reason.

surefooted : proceeding surely; not likely to stumble or fall.

symbiotic : pertaining to the closeness, proximity and affinity between two beings that are in mutual communication or maintaining mutually validating interactions.

sympathy : a sensation, feeling or emotion—of anger, fear, sorrow and/or pity—that is a *personal reaction* to the misfortune and failure of another being.

systematization : to arrange into systems; to systematize or make systematic.

systems theory : *see* general systematology

—T—

terminal (node) : a point, end or mass on a line; a point or connection for closing an electric circuit, such as a post on a battery terminating at each end of its own systematic function; any end point or 'termination' on a line; a point of connectivity with other points; in systems, any point which may be treated as a contact point of interaction; anything that may be distinguished as an 'is' and is therefore a 'termination point' of a system or along a flow-line which may interact with other related systems it shares a line with; a point of interaction with other points.

thought-experiment : from the German, *Gedankenexperiment*; logical *considerations* or mental models used to concisely visualize consequences (cause-effect sequences) within the context of an imaginary or hypothetical scenario; using faculties of the Mind's Eye to *Imagine* things accurately with *considerations* that *have not* already been consciously experienced in *beta-existence*.

thought-form : apparent *manifestation* or existential *realization* of *Thought-waves* as "solids" even when only apparent in Reality-agreements of the Observer; the treatment of *Thought-waves* as permanent *imprints* obscuring *Self-Honest Clarity* of *Awareness* when reinforced by emotional experience as actualized "thought-formed solids" ("*beliefs*") in the Mind.

thought-habit : reoccurring modes of thought or repeated "self-talk"; essentially "self-hypnosis" resulting in a certain state.

thought-wave or **wave-form** : a proactive *Self-directed action* or reactive-response *action* of *consciousness*; the *process* of *thinking* as demonstrated in *wave-form*; the *activity* of *Awareness* within the range of *thought vibrations/frequencies* on the existential *Life-continuum* or *ZU-line*.

threshold : a doorway, gate or entrance point; the degree to which something is to produce an effect within a certain state or condition; the point in which a condition changes from one to the next.

thwarted : to successfully oppose or prevent a purpose from actualizing.

tier : a series of rows or levels, one stacked immediately before or atop another.

time : observation of cycles in action; motion of a particle, energy or wave across space; intervals of action related to other intervals of action as observed in Awareness; a measurable wave-length or frequency in comparison to a static state; the consideration of variations in space.

timeline : plotting out history in a linear (line) model to indicate instances (experiences) or demonstrate changes in state (space) as measured over time; a singular conception of continuation of observed time as marked by event-intervals and changes in energy and matter across space.

tipping point : a definitive "point" when a series of small changes (to a system) are significant enough to be *realized* or *cause* a larger, more significant change; the critical "point" (in a system) beyond which a significant change takes place or is observed; the "point" at which changes that cross a specific "threshold" reach a noticeably new state or development.

transhumanism : concerning the next evolved state of the "Human Condition," which is to say either in a direction of "internal" or "spiritual" technologies that advance the *Self*, or the direction of "external" and "physical" technologies that either modify or eliminate the *Body*. In our present state of society, it is the "physical" that is selectively *sold* to the masses so that only a select few may experience the "former"; in *NexGen Systemology*, also referred to as "metahumanism" with an emphasis on "spiritual technologies" as opposed to "external" ones.

transmit : to send forth data along some line of communication.

traumatic encoding : information received when the sensory faculties of an organism are "shocked" into learning it as an "emotionally" encoded *"Imprint."*

turbulence : a quality or state of distortion or disturbance that creates irregularity of a flow or pattern; the quality or state of aberration on a line (such as ragged edges) or the emotional "turbulent feelings" attached to a particular flow or terminal node; a violent, haphazard or disharmonious commotion (such as in the ebb of gusts and lulls of wind action).

—U—

unconscious : a state when *Awareness* as *Self* is removed from the equation of *Life* experience, though it continues to be recorded in lower-level response mechanisms (fixed to a simulacrum or genetic vehicle) for later retrieval.

undefiled : to remain intact, untouched or unchanged; to be left in an original "virgin" state.

understanding : a clear 'A-for-A' duplication of a communication as 'knowledge', which may be comprehended and retained with its significance assigned in relation to other 'knowledge' treated as a 'significant understanding'; the "grade" or "level" that a knowledge base is collected and the manner in which the data is organized and evaluated.

—V—

validation : the reinforcement of agreements of Reality.

vantage : a point, place or position that offers a good view.

Venn diagram : a diagram for symbolic logic using circles to represent sets and their systematic relationship; named after the logician *John Venn.*

verbatim : precisely reproduced "word" for "word."

vibration : effects of motion or wave-frequency as applied to any system.

viewpoint : see *"point-of-view"* a.k.a. "POV" in *NexGen Systemology.*

vizier : a high ranking official; a minister-of-state.

—W—

wave-function collapse : see *"collapsing a wave."*

Western Civilization : the modern history, culture, ideals, values and technology, particularly of Europe and North America as distinguished by growing urbanization and industrialization and born from a rebellion to strong religious indoctrination.

will *or* **WILL** (5.0) : in *NexGen Systemology* (from the *Standard Model*)—the spiritual ability at (5.0) of an *Alpha Spirit* (7.0) to apply *intention* as "Cause" from a higher order of reasoning and consideration (6.0) than the thoughts found in *beta-existence*, where it manifests as "effect" below (4.0).

willingness : the ability and consideration to reach, face or confront some thing or energy; the ability and consideration to communicate along some line to produce an effect, to put attention or intention on the line.

—Z—

ziggurat : ancient Mesopotamian temples in the form of a stepped pyramidal tower presented as a series of seven tiers, levels or terraces.

ZU : the ancient cuneiform sign designating an archaic verb—*"to know," "knowingness"* or *"awareness"*; the active energy/matter of the "Spiritual Universe" (AN) that is experienced as *Lifeforce* or *consciousness* for entities existing in the "Physical Universe" (KI); *"Spiritual Life Energy"*; the spiritual energy present in the WILL of the actualized *Alpha-Spirit* in the "Spiritual Universe" (AN), which imbues its *Awareness* into the Physical Universe (KI), animating/controlling *Life* for its experience of *beta-existence* along an *Identity-continuum* called a *ZU-line.*

Zu-line : a spectrum of *Spiritual Life Energy (ZU)* as conceived on the Standard Model of Systemology; an energetic channel of *Identity-continuum* connecting the *Awareness (ZU)* of an *Alpha-Spirit* with *"Infinity"*; a *Life-line* on which *Awareness (ZU)* extends from the direction of the "Spiritual Universe" (AN) as its *alpha state* through an entire possible range of activity in its *beta state*, experienced as a *genetic-entity* occupying the *Physical Universe (KI)*; the Standard Model demonstrates the Zu-line interacting with spheres of existence.

Zu-Vision : the true Point-of-View (POV) perspective of the Self as Alpha-Spirit outside of the boundaries of the Mind-Systems and exterior to beta-existence.

REACH HIGHER.
JOIN THE ELITE.

SYSTEMOLOGY AIR COMMAND
TO EVOLVE - TO BECOME - TO DEFEND

mardukite.com

Humanity Needs YOU!

SYSTEMOLOGY
The Pathway to Self-Honesty

THE TABLETS OF DESTINY
Using Ancient Wisdom
to Unlock Human Potential
by Joshua Free

(*Mardukite Systemology Liber-One*)

A rediscovery of the original
System of perfecting the Human
Condition on a Pathway which
leads to Infinity.
Here is a new map on which
to chart the future
spiritual evolution of
all humanity!

CRYSTAL CLEAR
The Self-Actualization Manual
& Guide to Total Awareness
by Joshua Free

(*Mardukite Systemology Liber-2B*)

Take control of your destiny and
chart the first steps toward your
own spiritual evolution. Realize
new potentials for the Human
Condition in Self-Honesty.
Reclaim the freedom of the
spirit with a manual so profound
its effectiveness has been deemed
Crystal Clear!

THE SYSTEMOLOGY HANDBOOK: GRADE-III ANTHOLOGY

Unlocking True Power of the Human Spirit & The Highest State of Knowing and Being

This future may be shaped in Self-Honesty only by the highest caliber "spiritual technologies" at our disposal—methods that only a small esoteric demographic of the population has kept a possession of, and which is markedly only being developed, refined and radiated into this existence by those few who are "in the know."

In *"Tablets of Destiny: Using Ancient Wisdom to Unlock Human Potential,"* Joshua Free maps the highest route to esoteric knowledge—expertly bridging materials from the "Mardukite Research Library" with the wisdom and latest clarity revealed by the advanced "NexGen Systemology" New Thought division of the "Mardukite Research Organization." This provides a concise and accessible guide to the fundamentals of "cosmic ordering" and the "systematization" of the Human Condition, described on cuneiform tablets as "control of the Divine ME"—or else, the "Arts of Civilization"—the supreme knowledge and wisdom that commands all true authority of godhood in the heavens and sovereignty on earth. Access to this amazing potential is now available to all!

Take control of your destiny and chart the first steps toward your own personal spiritual evolution. Realize new potentials for the Human Condition in Self-Honesty. Reclaim the freedom of the spirit and achieve Ascension in this lifetime with a manual of techniques and teachings so profound, that its effectiveness has been deemed *"Crystal Clear."*

When a Seeker attains actualized Awareness outside of physical existence, therein alone lies the true personality of the Spirit, the individuated "I" that is Self. This is the state of Self free of worldly fragmentation. The Seeker has extended the reach of Awareness and the ability (and responsibility) of Self-direction from the point of WILL as Cause in the Universe, rising above lower planes of Effect and Desire.

The "Pathway to Self-Honesty" is what has uniquely defined the 21st Century paradigm of Mardukite Zuism and Mardukite Systemology—demonstrating its superiority as the preferred method of driving us forward into the future. Although it is our destiny to reach a superior state of Self-Actualization—to "walk among the gods"—the route or pathway that is our "fate" is ours alone to determine. Whatever detours and distractions we experience are entirely ours to decide—and the power and responsibility of such only thickens as we tread further and further on the Pathway to Self-Honesty.

Here is the map and compass to "pilot" your own destiny and the future evolution of Humanity!

For nearly a decade these matters of Mardukite Zuism remained privately in the domain of an advanced division of the "Mardukite Research Organization" known from its brief public references as the "NexGen Systemological Society" (NSS) governing the new "International School of Systemology" (ISS). This faction has operated in the esoteric underground since 2011, developing this direct extension of the "Mardukite Core" to support future research and discovery of continuing work, as it applies to the present state of the Human Condition and its imminent future.

Here are the Secrets of the Human Condition, Life, Reality and the Universe
known only to the most secret underground cabals throughout history and
the highest echelons of elite and Illuminati still alive and operating today!

MARDUKITE
ZUISM

THE JOSHUA FREE PUBLISHING IMPRINT
officially representing
MARDUKITE RESEARCH ORGANIZATION

MARDUKITE.COM